Studies in Big Data

Volume 100

Series Editor

Janusz Kacprzyk, Polish Academy of Sciences, Warsaw, Poland

The series "Studies in Big Data" (SBD) publishes new developments and advances in the various areas of Big Data- quickly and with a high quality. The intent is to cover the theory, research, development, and applications of Big Data, as embedded in the fields of engineering, computer science, physics, economics and life sciences. The books of the series refer to the analysis and understanding of large, complex, and/or distributed data sets generated from recent digital sources coming from sensors or other physical instruments as well as simulations, crowd sourcing, social networks or other internet transactions, such as emails or video click streams and other. The series contains monographs, lecture notes and edited volumes in Big Data spanning the areas of computational intelligence including neural networks, evolutionary computation, soft computing, fuzzy systems, as well as artificial intelligence, data mining, modern statistics and Operations research, as well as self-organizing systems. Of particular value to both the contributors and the readership are the short publication timeframe and the world-wide distribution, which enable both wide and rapid dissemination of research output.

The books of this series are reviewed in a single blind peer review process.

Indexed by SCOPUS, EI Compendex, SCIMAGO and zbMATH.

All books published in the series are submitted for consideration in Web of Science.

More information about this series at https://link.springer.com/bookseries/11970

Gitanjali Rahul Shinde · Soumi Majumder ·
Haribhau R. Bhapkar · Parikshit N. Mahalle

Quality of Work-Life During Pandemic

Data Analysis and Mathematical Modeling

Springer

Gitanjali Rahul Shinde
Department of Computer Engineering
Vishwakarma Institute of Information
Technology
Pune, India

Haribhau R. Bhapkar
MIT School of Engineering
MIT Art, Design and Technology
University
Pune, India

Soumi Majumder
Department of Business Administration
Vidyasagar University
Kolkata, India

Parikshit N. Mahalle
Department of Artificial Intelligence
and Data Science
Vishwakarma Institute of Information
Technology
Pune, India

ISSN 2197-6503 ISSN 2197-6511 (electronic)
Studies in Big Data
ISBN 978-981-16-7522-5 ISBN 978-981-16-7523-2 (eBook)
https://doi.org/10.1007/978-981-16-7523-2

This Springer imprint is published by the registered company Springer Nature Singapore Pte Ltd.
The registered company address is: 152 Beach Road, #21-01/04 Gateway East, Singapore 189721, Singapore

Preface

HAPPINESS is a state of mind that has nothing to do with the external world.

—Bhagwad Gita

This uncertain spread of COVID-19 is creating an emotional challenge for many people leading to their daily routine in very unprecedented ways. Every stakeholders of the society like employer and employees are being affected to large extent and are responsible to play important role in controlling this spread. Preventing and mitigating COVID-19 at the workplace are very important and require proper guidelines and recommendations to prevent and mitigate COVID-19 at the workplace. Work-life balance and mental health considerations are very important as per the guidelines issued by World Health Organization (WHO) during this COVID-19 outbreak. As per the guidelines stated, the recommendations are issues to the different groups of people which include the general population, healthcare workers, team leaders in healthcare organizations, older adults, people with underlying health conditions and their careers, people in isolation, etc.

Due to this pandemic, work from home culture has become a policy priority for most state and central governments and there should be policy in place considering both employers and employees. In view of this, an exploratory study is required to investigate the ongoing experiences of employees and employers faced in India. A question-based survey for employees in different sectors like education, information technology, agriculture, healthcare and real estate sector is required to collect their experiences on the different types of questions. On the responses received for these questions, mathematical tools can play an important role to build mathematical models for employees in different sectors. The application of appropriate tools from mathematics and mathematical model depends on the set of a questionnaire prepared for different domains. In addition to this, machine learning-based analysis is also important to draw inference from the outcomes resulted from clustering or classification algorithms in machine learning. The normative inferences are also one of the important outcomes of these machine learning-based inferences to decide what is ought to be? The selection of an appropriate machine learning algorithm will completely depend on the descriptive statistics of the data set, and accordingly, the outcomes can be analyzed for different purposes.

The key contributions presented in the scope of this book are listed as follows:

1. The preparation of questionnaires for employees in different sectors for recording their ongoing experiences during this COVID-19 pandemic.
2. Preparation of the data set resulted from the responses collected from the survey initiated for employees in different sectors.
3. A proposed sector-specific mathematical model to draw statistical inferences, correlation in the form of association rule mining and key observations.
4. Proposed machine learning-based analysis to draw useful inferences and normative interpretations from the data set
5. Set of useful recommendations and guidelines

The first part of this book explains the quality of work-life, its characteristics and stages along with the different factors causing its importance. Methodologies to improve the quality of work-life are also explained in detail in this part of the book. Mathematical modeling based on data analysis plays important role in predictions and forecasting. The next part of the book deals with the need for mathematical modeling, various tools and languages in mathematics addressed in this part. Pragmatic approaches for mathematical modeling based on rough sets, conditional probability and graph theory are explained in detail in this part.

The second part of the book covers specific case studies in various sectors which include real estate, education, software industries, and agriculture. This part presents and discusses an overview of every sector, the study of state of the art, descriptive statistics of the data set for qualitative and qualitative analysis. Mathematical modeling and machine learning-based analysis of the underlined data set is also presented in this part of the book.

The last part of the book addresses the need for principal component analysis (PCA) for categorical data set generated from the survey. In this part, an overview of PCA, computational stages and its interpretation is elucidated in this part of the book. In addition to this, the size of the sample data set, the margin of error, hypothesis testing, the challenge in data collection, analysis of the small data are also presented and discussed in the last part of this book.

The main characteristics of this book are:

- A detailed and in-depth description of all the topics related to all the topics.
- Use case and scenarios-based descriptions.
- Comprehensive mathematical modeling of the data sets pertaining to the different sectors.

- Machine learning-based analysis and normative inferences.
- Numerous examples, technical descriptions, program outputs and real-world scenarios.
- Simple and easy language so that it can be useful to a wide range of stakeholders like a layman to educate users, villages to metros and national to global levels.

Work-life balance and behavioral analysis of the data set for the purpose of mathematical modeling and machine learning-based inferences are very useful essentially in the scope of this uncertain pandemic. Because of this, this book is useful to all undergraduate studentsof these courses for project development and prediction/forecasting model design in machine learning and data modeling. This book is also useful to a wider range of researchers and design engineers who are concerned with exploring data science for recommendations and predictions. Essentially, this book is most useful to all entrepreneurs who are interested to start their start-ups in the field of data science and related product development. The book is useful for undergraduates, postgraduates, industry, researchers, and research scholars in ICT, and we are sure that this book will be well-received by all stakeholders.

Pune, India — Dr. Gitanjali Rahul Shinde
Kolkata, India — Soumi Majumder
Pune, India — Dr. Haribhau R. Bhapkar
Pune, India — Dr. Parikshit N. Mahalle

Acknowledgements

We would like to thank many people who encouraged and helped us in various ways throughout this book, namely our colleagues, friends and students. Special thanks to our family for their support and care.

We are thankful to BRACT's Vishwakarma Institute of Information Technology's trustees, director, all deans and head of the departments for their constant encouragement, inexplicable support.

We are thankful to Dr. Mangesh Karad, Executive President and Vice Chancellor, MITADT University, for his encouragement and constant support.

We would like to thank Dr. Nilanjan Dey Sir, JIS University, for his continuous support and constructive guidelines. We are also very much thankful to all our department colleagues at SKNCOE, for their continued support, help and keeping us smiling all the time.

Last but not least, our acknowledgements would remain incomplete if we do not thank the team of Springer Nature who supported us throughout the development of this book. It has been a pleasure to work with Springer Nature team, and we extend our special thanks to the entire team involved in the publication of this book.

Acknowledgments

Contents

About the Editors

Dr. Gitanjali Rahul Shinde has overall 13 years of experience, presently working as an Assistant Professor in the Department of Computer Engineering, Vishwakarma Institute of Information Technology, Pune, India. She has completed Ph.D. in Wireless Communication from CMI, Aalborg University, Copenhagen, Denmark on Research Problem Statement "Cluster Framework for Internet of People, Things and Services" in 2018. She has obtained M.E. (Computer Engineering) degree from the University of Pune, Pune in 2012 and a B.E. (Computer Engineering) degree from the University of Pune, Pune in 2006. She has received research funding for the project "Lightweight Group Authentication for IoT" by SPPU, Pune. She has presented a research article at World Wireless Research Forum (WWRF) meeting, Beijing China. She has published 50+ papers in National, International conferences and journals. She is the author of 5+ books with a publisher like Springer and CRC Taylor & Francis Group. She is also editor of books with De Gruyter and Springer. She is a reviewer of prominent journal IGI publications and IEEE Transactions.

Soumi Majumder completed her PGDM, specialization in Human Resource Management from All India Management Association (Institute of Business Management, Jadavpur University) in 2012. She did Diploma in Labour Laws with Administrative Laws (DLLAL) and Masters in Human Resource Management (MHRM) from Annamalai University, under Tamilnadu Government in 2013 and 2019, respectively. Currently, she is a Ph.D. student of the Department of Business Administration at Vidyasagar University, Midnapore, West Bengal, India. She is associated with the School of Business-Sister Nivedita University, WB, India, Department of Business Administration, Netaji Subhash Engineering College, Kolkata. Previously, she was associated with Techno India College of Technology, NSHM College of Management and Technology, Dinabandhu Andrews Institute of Technology and Management, J D Birla Institute of Science and Commerce, Future Institute of Engineering and Management, West Bengal State Labor Institute, Siliguri, Department of Management Science, Sister Nivedita University, Kolkata, India, etc. She is having 6 years of experience in academia and 2 years of industrial experience. She has 20 research papers in national, international conferences and journals in the area of quality work-life, decent work-life, work-life balance, stress management, employee engagement, job satisfaction, leadership, training, and learning, etc. She has presented a research paper at DSMLAI 2021 organized by India-Namibia Centre of Excellence in Information Technology and Namibia University of Science and Technology, Namibia. Furthermore, she is a member of the All India Management Association, ACM, IS, NIPM, etc.

Dr. Haribhau R. Bhapkar was graduated from University of Pune, India. He received masters in Mathematics in 2000, and Ph.D. (Graph Theory) in 2016. He is working as Professor in Mathematics at MIT Art, Design and Technology University‘s MIT School of Engineering, Pune. He has published 38 research papers in national and international journals and conferences, mostly in the areas of graph colouring, graph isomorphism, Mathematical Modelling and applications. He has developed his own academic credit system and defined a universal conversion formula to convert CGPA to an equivalent percentage. He has authored 141 books of Engineering Mathematics for different universities in India. He is the Research Guide in Mathematics at MIT Art, Design and Technology University, Loni Kalbhor, Pune. Additionally, he is the Board of Studies Member at MIT ADT University and Modern Arts, Commerce and Science College, Pune. His 21 years of teaching experience foster students on the path of successful personal growth and gratification. He is very keen to enhance the understanding level and problem-solving skills of students through his books and teaching. He motivates students to improve their concentration, innovative and design thinking, self-esteem and confidence to achieve desirable success.

Dr. Parikshit N. Mahalle obtained his B.E degree in Computer Science and Engineering from Sant Gadge Baba Amravati University, Amravati, India and M.E. degree in Computer Engineering from Savitribai Phule Pune University, Pune, India. He completed his Ph.D in Computer Science and Engineering specialization in Wireless Communication from Aalborg University, Aalborg, Denmark. He was Post Doc Researcher at CMI, Aalborg University, Copenhagen, Denmark. He worked as Professor and Head in the Department of Computer Engineering at STES's Smt. Kashibai Navale College of Engineering, Pune, India. Currently He is working as professor and head in Department of Artificial intelligence and Data Science, Vishwakarma Institute of Information Technology, Pune India. He has more than 20 years of teaching and research experience. He is a senior member IEEE, ACM member, Life member CSI and Life member ISTE. Also, he is a member of IEEE Transaction on Information Forensics and Security, IEEE Internet of Things Journal. He is a reviewer for IGI Global—International Journal of Rough Sets and Data Analysis (IJRSDA), Associate Editor for IGI Global—International Journal of Synthetic Emotions (IJSE), Inderscience International Journal of Grid and Utility Computing (IJGUC). He is a Member-Editorial Review Board for IGI Global—International Journal of Ambient Computing and Intelligence (IJACI). He is also working as an Associate Editor for IGI Global—International Journal of Synthetic Emotions (IJSE). He has published more than 150 research publications having 1711 citations and H index 18. He has 5 edited books to his credit by Springer and CRC Press. He has seven patents to his credit. He had worked as Chairman of various Board of Studies.

Chapter 1
Introduction

Abstract When it comes to imagining the role of workplace, COVID-19 has changed our perspective. In this new normal, employees are trying to adapt to new ways of working. The compromised productivity is coming out from these constantly blurring work-life boundaries. Eventual burnout, continuous distraction, overloaded of information, etc., are highlighted among employees. Ensuring effective quality of work-life, the balance between work and life must be the company's main focus. There is a need to understand the effect of pandemic in the Quality of Work-Life (QWL) of the employees. The following chapter showcases the characteristics, importance and effect of pandemic on QWL.

1.1 Introduction to Quality of Work-Life

In the business organizations, quality of work is a practice that enables at all levels (top level, middle level and lower level). This practice gives the organization an effective shape in terms of environment, methodologies and outcomes. The quality of work-life has twin goals to accomplish; it increases effectives of the functions of organizations. On the other hand, it helps to improve the quality of life of the working people at work [1]. In generic term, it can be said that quality of work-life is covering the area of people's feelings toward the dimensions of work in the work environment. This includes economical betterment like rewards and benefits, effective compensation structure, physical betterment like work conditions, health and safety at work, security and equity, physical condition of workplace, etc. Psychological betterment that considers under the dimension of quality work-life including appreciation, recognition, self-esteem, career development opportunity, training and development opportunity, job enrichment, etc. The growth and development of the organization is totally depends on how it works on attracting, recruiting, motivating and retaining its workforce. Nowadays, business becomes so much competitive; therefore, they are also focusing the talent of this manpower. They are planning on development of talent and competency level of the workforce and how it can be equipped with the going trend of the market [2]. If the people of the organizations are enjoying at their work, their outcome level will go beyond the standard performance. Their effective

G. R. Shinde et al., *Quality of Work-Life During Pandemic*, Studies in Big Data 100,
https://doi.org/10.1007/978-981-16-7523-2_1

performance makes the business performance efficient. The organizational objectives can be achieved easily by the entire firm with the help of human resources. The quality of work-life consists of various factors in the workplace, namely job satisfaction, stress reduction and management, cost-effective interventions and so on [2]. Therefore, the quality of work-life becomes very significance to run the business smoothly along with that the firm also can compete with the other competitors in the market. The revenue generation, reputation of the firm, recognition all are depending on the attractive and appropriate practice of qualitative work-life. To consider these phenomena, the bushiness house always need to formulate and maintain strategies to improve the employees' People will engage highly at their jobs and will produce high level of organizational commitment [3].

1.1.1 Characteristics of Quality of Work-Life

I. Employees' Attitude

Attitude of employees can be judged by their skills, knowledge, expertise, experience for performing his or her work. Every job position should contain particular attitude of the personnel like employees willpower to learn new things, effort for team work and openness toward creativity and innovation.

II. Fair Compensation Structure and Job Security

The business house should maintain an effective policy on compensation management. Salary and wage administration must be fair, equitable and justified to all. It also includes job security. If the compensation does not match with the performance level of employees, it will create feelings of dissatisfaction [4]. From the part of the employees', productivity level will be faced negative consequences. On the other hand, job security provides the assurance of permanent employment, and it generates improvement in the domain of quality of work-life [5].

III. Opportunities for Personal and Career development

Every organization should provide career development opportunities to create an atmosphere for better quality of work-life. In order to make the employees efficient at their job and cope up with higher job responsibility in future, proper training and development programs should be organized periodically [6].

IV. Balancing Between Professional and Personal Life

Apart from job duties and responsibilities, the organization should provide the time of relaxation to the employees' with their family members. To get higher productivity from the side of employees, balancing between professional and personal life is very much required. Work-life balance keeps them happy and satisfied [7]. They will not be overburdened and pressurized by huge work duties if proper balance between personal and work-life is followed.

V. Nature of Job

This refers to the allocation of job to the employees. If the job is dull, monotonous and routine kind in nature, then it will lead to boredom, dissatisfaction, fatigue. It makes quality of work-life poor [8, 9]. This can be modified by providing recognition, growth, creativity, advancement opportunities. In this way, the quality of work-life has a relation with nature of jobs.

VI. Level of Stress

Stress has negative impacts in productivity of an individual. The working people of the organization always need a stress free environment. If the stress level is high, the outcome declines, and if there is no stress, there is plentiful outcomes. The quality of work-life depends on stress free working environment of the organization [10]. The goals of the organization can be served in a better way.

VII. Risk and Reward

The individual should be paid more when they face risky and challenging job. Less risk and challenge gives them less amount of salary. For taking challenges in the job as well as risk factors, the organization must keep a well-defined reward system. It acts as a motivation to the employees. There is no doubt that highly motivated people will produce huge.

1.1.2 Stages of QWL Process

I. Flexibility in Work Schedule

Employees are the assets of the organization. They should be treated as resources. Therefore, a flexible work schedule must be maintained in the workplace. It includes a flexible working hour that is denoted by flexi time. Staggered working hours are also there. It means different time intervals between starting and ending working hours. Along with that compressed working week is organized where working hours are more per day, but numbers of working days are less in a week [11].

II. Autonomous Work Group

To create the positive feeling among the employees, autonomous work group should be created. Employees are given freedom in case of selecting their own teams. The freedom of decision making is also encouraged. The group members can design their own work schedule to accomplish the task. They have the right to choose the methods of work as well. It gives a boost in the area of practice of good quality of work-life.

III. Job Enrichment

Job enrichment makes the job very much challenging and interesting toward the employees. Therefore, there is no feeling of boredom, monotony and fatigue [12]. The degree of job enrichment helps to determine the degree of quality of work-life and level of job satisfaction.

IV. Opportunity for Development

An employee, who has higher career aspiration, always seeks better career development opportunity in the work place. If the job that has been provided to him by the organization is very enlarged and enriched, he or she will be feeling happy. They focus on the growth of the career. When personal development is there, it leads to personality development. Good quality of work-life provides high level of employee engagement. Employee engagement creates a high level of commitment.

V. Stability of Employment

The stability of employment is very important component under this juncture. Every employee wants stable employment conditions [13]. Job security increases the performance level of the employees. The summations of employees' performance help to make overall organizational performance efficient. The business can perform better and easily compete with the competitors in the market.

VI. Participation

Participation of employees in management's decision-making process makes the employees accountable. When they are participated and their views are appreciated and valued, then it provides happiness to them. They are very much motivated by this and feel a strong bonding with their workplace. Employees' participation in management also improves the quality of work-life in a better way.

VII. Recognition

Financial remuneration is not enough to boost employees' performance. To motivate them, highly organization needs to establish non-financial benefits like offering membership in clubs or other associations, offering vocational trip with family, job enrichment opportunity, award and reward ceremony, well-furnished workplaces with decent work culture and many more.

VIII. Harmonic and Congenial Relationship

Harmonious and congenial relationship between management and employees can make a sense of belongingness. Good industrial relation reduces the chance of industrial disputes. The various forms of disputes like strikes and lockout must be eliminated by the process of production and business functions as well. Every business should do smooth functioning and make the wheel of production continue. Management and workers relations are very crucial to improve the quality of work-life [14]. In a team work, the relationship between superiors and subordinates also should be congenial.

IX. Grievance Handling Procedure

Grievance is the dissatisfaction feeling of employees regarding organizational factors. Grievance creates low quality and quantity of production, increases the wastage of materials, spoils the leakages of machineries and tools, increases absenteeism and turnover, decreases the commitment level. Therefore, grievance must be handled by the management of the organization. If any employee faces problems related to work factors, he or she must expresses grievance to the concerned authority. Grievance management is one of the most effective practices of organization. By doing counseling and other methods, management can abolish employees' grievance. Effective grievance handling procedure provides quality work-life to employees.

X. Occupational Stress

The HR managers of the organization identify the stress level. Stress is not at all good for employees. It gives pressures to physical and mental state of the employees. When there is a gap between the expectations and reality in job demands for the manpower, it is creating stress toward the job. It increases when employees are not able to meet the job demands. The results of stress is swing the mood, low energy level, stomach upset, frequently caught by cold and infections, etc. Employees are the critical resources of the organizations [15]. If they are healthy, they are productive. Productive employees bring organizational success.

1.1.3 Importance of Quality of Work-Life

Various research studies and theoretical models have been established that the retention level of the working class in an organization can be increased by focusing their needs and demands. In this regard, the organization can be benefitted in the area of productivity, employee loyalty, recognition and reputation in long term. Organizational effectiveness is highly influenced by quality of work-life [16]. It has been said that it is the input of level of employee satisfaction and result is increased economic strength for the organization. There are so many significant factors that make quality of work-life very important in the paradigm of business environment. This helps (i) to raise productivity and profitability of business, (ii) it attracts talent and retain them for long run, (iii) the rate of turnover and absenteeism is minimized, (iv) the work-life quality is continuously improving, (v) a proper work-life balance provides benefits to the families and communities (vi) generates job satisfaction and finally (vii) builds positive image of the organization.

The efficiency level of employees indicates rising productivity. Work–life balance creates influence on profitability index of the organization. On the other hand stressful environment of work effects negatively on productivity. Therefore, company's management needs to form such policies, values, norms, ethics that show a positive image in the field of business effectiveness. This will surely help the employees to engage entirely from their side. The loyalty of the employees must be created by providing a good quality of work-life. Through the research works, several new strategies are implementing in area of business, and it becomes easier to understand how to attract and retain skilled and knowledgeable personnel. Attractive and suitable policies framed by the top-level management always focus on the practice by which the people can stick for a longer period of time. Employees are full of human capital, and they are motivated at their performance by receiving performance related incentives, leave with pay facility, comfortable hours of work, recess or break during job hours, career development opportunity, appreciation, advancement and growth, etc. In the process of managing employees, there is an involvement of grievance, dissatisfaction at job, conflict with management, indisciplined behavior, misconduct by the side of the working people. If these negative components are not handled properly by the management, they will face consequences that are not good for work environment. The name and image of the company also will be hampered [17]. By introducing qualitative work culture in the workplace, the system can reduce and omit these problems. Work-life balance is minimizing fatigue, monotony, boredom and stress. The responsibility of the employers is to provide and support the community activities and help the employees to engage there. In this way, the community and society at large must be benefitted by employers. Quality of work-life is making an increased level of job satisfaction. Satisfied employees are loyal and committed toward the organization. For the employers, the quality of work-life phenomena helps to create a benchmark in the market and society for the employers.

1.1.4 Methodology to Improve QWL

I. Reward and Recognition

In the organization, there should be established a reward and recognition system that can motivate employees at their performance. The motivated employees are contributing a lot in terms of energy and time. Compensation is not enough to boost the employees' efficiency level at their job, every employee wants something above this remuneration, and they want appreciation and always expect that their work will be recognized by higher management.

II. Growth Opportunities

The other way to improve quality of work-life of employees is when they receive better career development opportunities.' Personal growth helps to achieve organizational goals. Growth and development satisfy the self-actualization needs of the employees. It is the desire of individual. Everyone has high or low career aspiration. Growth is the inborn opportunities of human beings. The growth and development opportunity is based on promotion, adding value to jobs, constructive and worthy job, providing learning opportunity, job enrichment, job enlargement, advancement in career, etc. [18].

III. Respect and Self-esteem

In any situation into the organization, as people are resources, therefore equal treatment should be given to them. Equality in treatment gives them the sense of belonging to the organization. Discrimination, harassment, victimization are the reasons of poor quality of work-life. Managers should show respect to all levels. The self-esteem needs help people to enhance and improve. People want to be recognized and appreciated by their work. Authority, status, power, fame, etc., make them to feel happy and highly engaged. An example can be written based on a pacemaker manufacturing company (1989); they initiated a program of organizational improvement through quality team work. In this program, it has been said by the general manager of the company to the workers that they are paid for their two hands. As they are using the traditional method of work while working, but they never put any suggestions to improve the level of business performance. In the meeting, they were asked to use their brain, the excellent asset as per the GM. The company will be benefitted highly, along with the two hands if these people use their brain. They should apply new ideas, views and suggestions' to enhance the level of overall performance. In this way, the workers felt they are respected and treated better. The result is high engagement and commitment to make the project successful. The company achieved outstanding results and revenue. If people are feeling respected, they give their best to attain the goal at any cost.

IV. Support and Vision

Unless the people know the meaning of commitment, they cannot make any commitment. The vision of the company should be clearly communicated to the employees of the organization. People should know exactly what is expected from them. When they get supports from leaders and vision is very vivid to them, they can do work accordingly. An effective leader must support, guide, direct and motivate people to accomplish the vision statement of the organization. The transformation of the organization is very important scenario and can be made by the joint effort of leaders and followers. People also have some career aspiration. By the support of their immediate superior, they can increase their personal development and organizational development. Management by objectives or MBO technique that is established by Peter Drucker, one of the famous Management scientists, had said in the process, that objectives can be achieved by the mutual effort of superiors and subordinates.

V. Suitable Working Conditions

In the present scenario of business, the people are spending most of the time in the work place. A congenial work environment is highly appreciated by the working people. They want to spend happy and quality time to do work together. Effective team work, cooperation, coordination, good interpersonal relationship between employees create a good working environment. This attracts employees and also makes a long retention level with the company. On the other hand, politics, back biting, rumors, etc., give a negative impact on the working environment. It discourages the employees to do the work. The quality of work-life is based on suitable working environment like the other factors.

VI. Empowerment

Empowerment means distribution of power, authority, freedom and responsibility among the employees by employers. The people are feeling that their work is worthy when they are empowered. A sense of control over their work always presents while they are working. A greater job satisfaction has been generated by this practice, empowerment. Job satisfaction leads to better commitment and loyalty. Empowerment is the motivational tool that helps to grow independency among them.

VII. Work-Life Balance

Quality of work-life is also measured on the basis of work-life balance. This is the phenomena that saying about the management of professional life and personal life of the employees. This encourages people to stay long run in the organization. To make quality of work-life effective, work-life balance is the main concern of the company. More they become happy, more they productive. Too much work pressure give stress to the employees. Stress is not a good sign [19]. It develops frustration among employees. Mental and physical health both will be hampered by excessive stress. Therefore, employees should spend some quality time with family members, relatives and friends. To keep in mind, organization must play roles to maintain a balanced work-life among personnel.

1.2 Effects of Pandemic on Quality of Work-Life of Employees

Quality of work-life is a prime practice in any organization under the vast field of human resource management. It is very much significant and has a huge influence on the people who are engaged in different establishments. The quality of work-life indicates the employees' statues in relation to their happiness and level of job satisfaction. If we focus on one of the important parameter of qualitative work-life, we must consider work-life balance. In the UK, the term work-life balance was first introduced in the women's Liberalization Movement. It was said that time that average working hour should be 14 to 16 h in a day and in a week it should be six days [20]. In this way, the entire concept of quality work-life became very popular in the reform and re-establishment strategy. Quality of work-life provides employee engagement in the organization. High level of engagement leads to high level of commitment. Committed people produce more. Out of the productivity, the business organization can achieve its objectives. In present time, due to the drastic effects of pandemic, the quality of work-life of the working people is getting poor day by day. Employees are far away from the workplace and are doing remote work in this scenario. All the time, all kinds of facilities and resources are not available to do work through this man–machine interface [21]. The work from home concept makes the work-life balance unbalanced. Working people are not able to manage professional life as well as personal life during these phases as lots of work are coming to them from their workplace as well as from home too [22]. Due to the enhancement of technology, the employers are shifted to run their business operations through work from home method. The lives of the people are drastically changing for the wildfire spread of COVID-19 pandemic. It gives the adverse effect to the young people's lives as well. Schools and universities have been closed, exam and other events most of the time got postponed, and socializing with friends and family also is stopped

[23]. It is getting tough for the people including young to live in this circumstance. Physical and mental wellbeing is getting hampered. Not only the young are suffering in this way, the working class of the country is also facing negative consequences. They are losing their quality of work-life. To keep in mind the effects of pandemic, business houses are trying to do their operations at their best, but revenue generation or profit maximization are not fulfilled by most of the organization in this ongoing situation. Therefore, termination, attrition, lay-off, pay cuts, etc., are taking place. People become very much unsecured in their job. They are facing job insecurity at work. As compensation is most important factor for all the job holders, this is their bread earning and sustainability. Being faced several uncertainties in the pandemic, people are getting afraid off lose the job. In spite of knowing the fact of overburden of work, they are not able to do anything opposite of employers; rather, they are adapting the changes with new flow of work and the work culture in new normal. The career development opportunity, training and development facility, monetary and non-monetary facility and other welfare facility all become nil in most of the organizations [24]. This makes poor quality of work-life for employees.

The work-life balance indicates the picture where the employees can manage their professional life as well as personal life too. In this pandemic time, work from home practice becomes easier due to availability of cloud based software, mobile technology, advantages of smartphone and prolong activities of the Internet. A prime feature of poor quality of work-life is stress. It can be said that by using these technologies the business entrepreneurs can run their operations easily, but on the other hand, the excessive level of stress makes the employees life blur. It becomes such a practice where employees need to produce more in less time; result is physical, and mental health problems are raised. Another striking thing in this crisis time; new trend of work has been observed and named as 'bossware'. It is one type of controversial software implemented by business organizations. Through this software the work of the employees can be tracked and monitored, while they are working from home. It is the interpretation of the management to enhance the productivity of business by this process. People are getting afraid due to under surveillance. The outcome is long working hours for the employees [25, 26]. It is observed that proper tea break, lunch break, attending invitation to friends' home all are going to be vanished. Employers' expectation is reaching beyond the level, and gradually the working hours take place like 24 $\times$ 7. This is very much true that employees are the crucial assets of any business house; by making them happy and satisfied, the organization can sustain in the market for long run to create a positive image of the firm. To generate a better quality of work-life, this is the responsibility of the employers to make a stress-free environment at work. There is no doubt that quality of work-life becomes worst, and in the context of human resource management, it has no justification. Every organization should follow a suitable work environment and establish transparent work-related policies that will encourage people to do work with happiness. Staffs must be provided breaks and leisure time for spending with family members. Several research studies have shown that large working hours affects health of individuals. It creates cardiovascular disease, and due to mental health issue, people work-life can be traumatized. Poor work-life balance is the reason of sleepless. Sleep deprivation is very much connected with cardiovascular disease. This also effects on physical health. According to research studies, normal sleeping schedule should be on an average 7 to 9 h per night. Due to these negative effects of stress, there is a high chance of stroke, coronary heart disease and sometime mental disorders such as depression, frustration and anxiety. Myocardial infarction and stroke- are causing death. The two most significant factors are considered here to make non-effective quality of work-life; they are sleep deprivation and myocardial infarction [27]. The need of work-life balance is more important than ever, in this pandemic situation.

COVID-19 pandemic has got the birth of unpredictable work environment including pressured work culture for the employees. During the period of isolation at home, it becomes difficult for them to face the physical and psychological challenges. In order to maintain a good and effective quality of work-life, not only the employees are responsible, also the employers should extend their hand of help toward the employees. They should provide all the welfare facilities and betterment that are really needed in this time. The employees are the human capital of the organization. By using this human capital, the organizations can enable to generate financial assets in true sense. Pandemic effects on employees are really horrible. Their lives are facing risk also [28]. To keep in mind the importance of human assets, the system must implement necessary attributes of quality of work-life.

1.3 Strategies to Reduce the Pandemic Impact on Employees

Over the last few months, it has been seen that due to COVID-19 virus many people are getting infected in many of the countries all over globe. With the speedy spread of this corona virus, a negative impact has come in the area of economy, society and business as well. During this pandemic, the business leaders have become very much concerned to reduce the horrible impact of corona virus on business and employees and their level of performances. According to Deloitte survey, it has been stated that the pandemic has created an adverse effect on sales volume and cash flows. It is also creating some risk factors including lack of supply of raw materials, lack of tools, machineries to execute remote work, safety issues of employees and so on. The executives of the business firms also mentioned that COVID-19 has made a drastic change in the fields of organizational efficiency [29, 30]. The performance target has been decreased by the employees of the firms. On the basis of Deloitte research work, some interventions are made as a reduction strategy of the pandemic impact on employees.' This is a universal truth that employees are the assets of any organization. By making them productive in the organization the business can earn a higher level of profitability and improves, the efficiency level of its performance.

1.3.1 *From Company's Perspectives*

In this pandemic situation to decrease the negative impact, the companies should establish a cross-functional team, where all the team members are very much dedicated toward accomplishment of objectives. To coordinate the activities of different business units, the cross-functional team should provide necessary information related to tasks to senior management. Every operations of different business units must be monitored by this cross-functional team [31]. After the senior management, the necessary information can be communicated with the employees of the firms, customers, shareholders and partners. It is the peak time to analyze the critical roles, responsibilities and key job positions in the organization. In this time, it may happen that top management is away on business trips and also shortage of some efficient employees due to the reason of sickness or in quarantine. This does not mean that work stoppages are there; rather, the companies are developing a suitable and appropriate process of management regarding decision making under different scenarios of business. The company needs to make safety guidelines to them who are at workplace and cannot able to work remotely [32]. A periodical review is needed to maintain proper hygiene in the workplace with applying disinfectants. The management of the company must focus on analyzing the priority of the projects. The most important and critical tasks of its projects and efforts of the teams should be nurtured in this case. Low-priority tasks are given high flexibility.

The continuous communication between leaders and companies' people during this critical time can leave a lasting mark forever. An effective communication and interaction among them make the company powerful and rejuvenate its work culture. It is established that if there is a disruption, there always be a recovery. Company must facilitate some remote work options that are flexible in nature. The business should consider making a temporary basis succession planning for the company's key executive positions and critical role positions. Always share the symptoms of COVID-19 disease and the prevention mechanisms among the employees. Conducting a series of remote seminars and workshops with expert or health professionals is also required very much to create awareness about the severity of the pandemic. In this sessions, the question–answer round is also encouraged by these experts. They must try to give all the answers that are asked by employees. During this critical time, the employers also consider to provide financial and psychological support to the manpower. They are providing additional insurance coverage, assistance in emergency, payroll payment regular wise. Medical tools, equipments like thermometers, antibacterial products, etc., are purchased by the management of companies to maintain proper health and safety norms in the workplace [33]. They are very much concerned about to monitor employees' health, more over always focused on disinfecting the work environment. Apart from activities, the employers need to review the existing sick leave policies for physical betterment of employees. If there is a rise of illness among employees, they do not need to show medical documents all the time. There leaves or absences are to be considered on the medical ground.

Business organizations should maintain proper health and safety in workplace. By doing disinfecting and cleaning throughout, safety can be maintained. Any suspect of being effected by COVID-19, the employer needs to remove the employee from the entire work and send him or her for treatment. The organizations with high travel need make sure that transmission rate should be minimized during traveling to international destination. For this purpose, all types of safety measured must be taken by company. Sometimes, companies make mandatory rules on 14 days isolation who have come from abroad. There is a need to monitor the latest travel guidance from the government actively, reviewing travel policies, and on this basis preparing a track report and communicating the employees who are travelers in this scenario. Due to an adverse impact of this pandemic, internal employees should follow all the guidelines in order to meet the external clients and customers. Policy on social media should be very much active [21]. The monitoring program on social media may help identifying emerging issues of the pandemic and its horrible effects on markets and regions of production. Employees are updated about the challenges that have been created by this crisis situation periodically. Employers can take an initiative to generate their own news channel in the workplace on the basis of authentic and credible sources of information such as information published from World Health Organization. At the end of the day, it is very much true that human resources are the critical assets of the organization. Anyone can be infected by the COVID-19, and therefore, the prime responsibility of business leaders is to lead for the physical, mental betterment of the people and provide them optimum level of welfare facilities to make them healthy and fit [34].

1.4 Work-Life Balance: Tips to Reduce Pandemic Pressure

The people of India are not used to do work through the new method 'work from home.' It is one kind of fancy idea of doing work. In this way, working people can do work from anywhere and also spend time with family, relatives and friends. Remote working gives negative impact on wellbeing of people. It makes people stressed. The working hours become 24 × 7 during this ongoing pandemic situation. People are getting frustrated as they have to meet high targets at work. Digital communication has taken an effective role in this scenario to maintain all kinds of professional communication. Study has revealed that Indians have faced a high level of burnout over the last six months. As per Microsoft Work Trend Index, it has been said that India has got second rank in case of highest number of workers who faced burnout in Asia, and the figure is 29%. Statistics has shown that remote working practice creates an additional 60 h of work in a month for the working class due to this COVID-19 [35, 36].

COVID-19 has changed our perspective when it is coming to the envisioning role of the workplace. In this new normal, the employees have tried to adapt new ways of doing work. The compromised productivity is coming out from these constantly blurring work-life boundaries. Eventual burnout, continuous distraction, overloaded of information, etc., are highlighted among employees. Many companies are worried to give calls for their employees for rejoin due to fear of speedy transmission of infection. The work-life balance of the employees is getting poor. The situation has made adverse impact on mental as well as physical heath of the working people. Few important tips are jotted down to maintain a perfect work-life balance [37].

1.4.1 Reduce Multitasking

Several research studies have shown that in this phase multitasking can hamper productivity of the work. This creates poor quality of work-life. It is suggested that people should target on one thing at a time. Continuous correspondence through emails and messages should be reduced. They need to stop themselves for being responsive always and should create a boundary to eliminate conflict.

1.4.2 Deep Breath and Meditation

Meditation helps to reduce job-related stress. People who are suffering in high stress, depression, frustration, they should do meditation regular basis. The deep breathing technique also helps to create relaxation. It improves and controls the ability of reaction to negative feedback. To diminish the stressors, people should hold the breath for 7 s and exhale for 8 s. This pattern of breathing is aiming to decrease anxiety and inducing sleep [38].

1.4.3 Exercise and Eating

Physical and mental health's are very much related to each other. Due to excessive stress, not only physical health has been hampered, the mental health also is suffered. Brain is the assets of manpower. They are using it in the workplace and making high level of productivity. When it comes to stress and burnout, these brains become brain-drains. Not only the employees, but also organizations face loss due to poor work-life balance of employees. To overcome this, one should consume rich foods and maintain healthy diets. Rich foods consist of fresh fruits, vegetables and whole grains that have the ability to improve immune system and fight against deficiency. Along with this, regular exercise is very much needed. Headaches, fatigue, anxiety, stress, etc., will be removed through the practice of exercise and good foods. Regular exercise releases endorphins, and it creates stabilized mood, helps to improve sleep and increase the power of self-esteem to people.

1.4.4 *Define Individual Space at Home*

In the present time, as work from home practice is ongoing for all the establishments, everyone needs to select a proper place while working. Otherwise if anyone does his work to stay with family members, there are several kinds of responsibility that can distract the person from work [39]. Moreover, household duties also can come while working. This also gives stresses to employees. Therefore employees need to always choose a dedicated workplace at home. When they have meeting to attend, they should use this remote place at home. It also helps to maintain a balanced work-life.

1.5 Employee Engagement Practices During COVID-19

1.5.1 *Virtual Yoga or Workout*

For elevating mood of the employees to get a little blood pumping means a lot. To be physically and mentally fit, employees need to take a quick pause from work to attend the session of virtual yoga or workout class. Team leader can join the class with the team members that are run by an expert [40]. Due to this work from home practice, people are getting stressed and sometime depressed. This workout or yoga will help them to engage at work actively.

1.5.2 *Online Happy Hour*

Employee engagement does not mean that people always need to stay at work. The team members are cooperated to make the operations effective of a team, but sometime, they should know each other personally not only through work. Therefore, casual settings of work also make the employees engaged. They will know better their professional role in this way. The management can use their favorite video conferencing platform like Zoom, Gmeet or Microsoft Teams to enjoy beverages together. To keep in mind, the drastic effects of the pandemic fancy cocktail party through virtual mode may be arranged to make the employees happy and engaged in work spontaneously. Some lighthearted fun is also included under this juncture.

1.5.3 *Cook Together Online*

While working at home, the employees also can share few recipes with others through the virtual platform. It is also one type of mind refreshment practice when everyone is showing the recipes of their favorite meal. Budget of some grocery items can be shared for cooking purposes. It should make sure that all the meals prepared by the ingredients maintaining all dietary restrictions, so nobody can be left the meeting [41, 42].

1.5.4 Virtual Team Building Activities

During this pandemic, the work-life along with the personal life has been hampered. The morale and job satisfaction level is very poor for the employees, and it has created an adverse effect on quality of work-life. The employee engagement level faces some negative consequences that is why commitment level toward the organization is also suffering. People should be cherished all the time to be happy and work effectively. Their mental health and physical health should be kept in a safe mode. In this area of discussion, virtual team building activities are useful tool to engage employees and make them connected with others while social distancing is on. Virtual team building activities include a lot of things of fun, namely clue murder mystery, team pursuit, code break, game show extravaganza, virtual beach staycation, virtual escape room, virtual office Olympics, virtual time machine, etc. Clue murder mystery is a fun activity where team members are solving who had the motive or commit a crime. Team pursuit is making the team strong. The team members can enhance their mental skill, physical skill and mastery. Code breaks give the opportunity to the group members to solve puzzles, riddles in a competitive environment. Game show extravaganza is related to the questions on pop culture, sports, world history, geography and more. Virtual time machine transforms home offices into a virtual game show experience including scoreboard, buzzer and all. Virtual beach staycation is a team building adventure where taste and flavors of vacation can be taken by the group. Escape room is the activity which is based on robbery. It is a brain storming session of group members. The mastermind of the group will steal precious jewels, and it is the duty of other members of the group to recover them before the time goes out. Some hidden clues are there; people need to think deeply and should solve the problems. This activity requires high collaboration and creative mind for problem solving. Virtual office Olympic is also a team building activity where the teams are competing with each other in relation to finish their assignments or tasks, which group can take the chance or challenge to present their teamwork in front of others first [43–45].

1.5.5 Games on House Party

House party is a type of app that is connected with face-to-face social networking. Over the last few months, this app becomes very much popular toward one of the useful modes of employee engagement index. This is well known in abroad rather than in our country. House party comes into action with some new and unique social features, which are unique games. Employees can spend some quality time with each other with the app. The app allows many numbers of available games; by this, employees can connect with their co-workers. The names of the games are like heads up, trivia games, etc.

1.5.6 Virtual Group Training Programs

In this pandemic time when employee morale is very much down all over globe, there is something that can be considered as teams' personal and professional development. Training is the process by which employee can enhance their skills and knowledge. Training is short term where development is long term in nature. To boost up the moral of the employees, few effective training should be provided to them such as emotional intelligence or EQ-based training, time management training on practical basis, effective communication and training. EQ training helps to develop leadership, self-awareness and social skills. Practical time management training provides the concept of minimum distraction while working and make the time optimum utilized while working at home and in the office both. By the help of effective communication training, the team members are becoming efficient communicators. This helps to develop verbal, non-verbal and written skills of the employees [46, 47]. People can be engaged in this way, and they can improve their physical and mental betterment.

1.5.7 Host Virtual Lunch and Learning

Another practice of employee engagement in this pandemic time is to celebrate virtual lunch with team members along with learning. A time should be fixed for this when they can make some research work to their relevant industry as well as their assignment too. The findings can be presented to the teams; this method is very useful and provides deep interest to working class [48]. Once in a week, this activity can be arranged. Being connected to the rest of the team through the electronic mode of communication, key soft skill learning and practice is easier for all.

1.6 Discussion

COVID-19 outbreak has presented unprecedented circumstances before education, real estate, health, agriculture sectors, etc. The lockdown has been announced to control the spread of COVID-19 infections; however, people/industries/organizations were not ready for lockdown, and it is affected greatly on their growth. The front workers in the healthcare sector were suffered a lot as major responsibilities they needed to carry on. The education sector is also hampered due to the pandemic as schools, colleges were closed and teaching, examinations were carried out on online platforms. These platforms were new to teachers as well as students. The real estate sector faced tremendous loss in this pandemic as people were scared and no one was ready to invest their money in such an uncertain time. The agriculture filed is also suffered as raw materials required for agriculture were not available readily due to pandemics. There is a need to understand and analyze the impact of the COVID-19 pandemic on the Quality of Work-Life (QWL) in these mentioned sectors. Various mathematical modeling besides machine learning techniques may help to understand and draw useful inferences and normative interpretations.

1.7 Summary

It is well established from the period of Taylor's scientific management (1910) system that management should take care of the betterment of the employees. This is the duty of employers to improve the quality of work-life, not only this pandemic time rather this should be a continuous process to make employees' productive. Better quality of work-life ensures greater participation and higher involvement of employees. Quality of work-life directly and indirectly connected withthe retention ratio of skilled, talented manpower in the organization. Due to poor qualitative work-life, the talent can move from one organization to another organization, and this is a loss for the organization as there is involvement of high replacement cost. In this new normal time though the quality of work-life and work-life balance have been destroyed, but this is not a good sign for them as it creates lots of problems in relation to their mental and physical health. Human resource plays the key roles such as developing, reinforcing and changing the culture of the organization. Therefore, their quality of work-life must be taken care. Basically, quality of work-life covers all essential elements of employees as well as business. This is based on recruitment and on boarding of employees, retention strategy, managing employees performance, training and development program, pay and perks management, discipline management, grievance handling procedure, development program for executives and many more. Manpower is highly motivated when they receive these facilities from their employers. The brand value of an organization not only depends on effective usage of physical, financial and technological resources, also highly focused on human resources and best practice on quality of work-life to make them productive continuously. This helps to improve the functions of business in its all sphere.

References

1. Saraji, G.N., Dargahi, H.: Study of quality of work life (QWL). Iran. J. Public Health **35**(4), 8–14 (2006)
2. Nadler, D. A., & Lawler, E. E. (1983). Quality of work life: perspectives and directions. Organ. Dyn.
3. Lawler, E.E.: Strategies for improving the quality of work life. Am. Psychol. **37**(5), 486 (1982)
4. Cummings, T.G., Molloy, E.S.: Improving productivity and the quality of work life. Praeger (1977)
5. Sirgy, M.J., Efraty, D., Siegel, P., Lee, D.J.: A new measure of quality of work life (QWL) based on need satisfaction and spillover theories. Soc. Indic. Res. **55**(3), 241–302 (2001)
6. Havlovic, S.J.: Quality of work life and human resource outcomes. Ind Relat: J. Econ. Soc. **30**(3), 469–479 (1991)
7. Kraut, R.E., Dumais, S.T., Koch, S.: Computerization, productivity, and quality of work-life. Commun. ACM **32**(2), 220–238 (1989)
8. Adhikari, D.R., Gautam, D.K.: Labor legislations for improving quality of work life in Nepal. Int. J. Law Manag. **52**(1), 40–53 (2010)
9. Kanten, S., Sadullah, O.: An empirical research on relationship quality of work life and work engagement. Procedia Soc. Behav. Sci. **62**, 360–366 (2012)
10. Fields, M.W., Thacker, J.W.: Influence of quality of work life on company and union commitment. Acad. Manag. J. **35**(2), 439–450 (1992)
11. Chan, K.W., Wyatt, T.A.: Quality of work life: a study of employees in Shanghai. China. Asia Pacific Bus. Rev. **13**(4), 501–517 (2007)
12. Crompton, R., Lyonette, C.: Work-life 'balance' in Europe. Acta. Sociologica. **49**(4), 379–393 (2006)
13. Lockwood, N. R. (2003). Work/life balance. Challenges and solutions, SHRM research, USA, pp. 2–10.
14. Delecta, P.: Work life balance. Int. J. Curr. Res. **3**(4), 186–189 (2011)
15. Kalliath, T., Brough, P.: Work–life balance: a review of the meaning of the balance construct. J. Manag. Organ. **14**(3), 323–327 (2008)
16. Fleetwood, S.: Why work–life balance now? Int. J. Hum. Res. Manag. **18**(3), 387–400 (2007)
17. Lewis, S., Gambles, R., Rapoport, R.: The constraints of a 'work–life balance'approach: an international perspective. Int. J. Hum. Res. Manag. **18**(3), 360–373 (2007)
18. Gregory, A., & Milner, S. E. (2009). Work-life balance: a matter of choice? Gender. Work Organ. **16**(1), 1–13.
19. Majumder, S., & Biswas, D. (2021, February). Fundamentals, present and future perspectives of quality of work life. In: Proceeding of first doctoral symposium on natural computing research: DSNCR 2020 (Vol. 169, p. 271). Springer Nature.
20. Majumder, S., & Biswas, D. (2021). COVID-19 impacts construction industry: now, then and future. In: COVID-19: prediction, decision-making, and its impacts (pp. 115–125). Springer, Singapore.
21. Schieman, S., Badawy, P.J., Milkie, M.A., Bierman, A.: Work-life conflict during the COVID-19 pandemic. Socius **7**, 2378023120982856 (2021)
22. Duan, Y., Iaconi, A., Song, Y., Norton, P.G., Squires, J.E., Keefe, J., Estabrooks, C.A.: Care aides working multiple jobs: considerations for staffing policies in long-term care homes during and after the COVID-19 pandemic. J. Am. Med. Dir. Assoc. **21**(10), 1390 (2020)
23. Dewey, C., Hingle, S., Goelz, E., Linzer, M.: Supporting clinicians during the COVID-19 pandemic. Ann. Intern. Med. **174**, 752–753 (2020)
24. Jaiswal, A., Singh, T., Arya, Y.K.: "Psychological antibodies" to safeguard frontline healthcare warriors mental health against COVID-19 pandemic-related psychopathology. Front. Psych. **11**, 1415 (2020)
25. Li, J., Ghosh, R., & Nachmias, S. (2020). A special issue on the impact of the COVID-19 pandemic on work, worker, and workplace!? Implications for HRD research and practices in time of crisis.

26. Creese, J., Byrne, J.P., Conway, E., Barrett, E., Prihodova, L., Humphries, N.: "We all really need to just take a breath": composite narratives of hospital doctors' well-being during the COVID-19 pandemic. Int. J. Environ. Res. Public Health **18**(4), 2051 (2021)
27. Palumbo, R.: Let me go to the office! An investigation into the side effects of working from home on work-life balance. Int. J. Public Sect. Manag. **33**(6/7), 771–790 (2020)
28. [30] Amin, K. P., Griffiths, M. D., & Dsouza, D. D. (2020). Online gaming during the COVID-19 pandemic in India: strategies for work-life balance. Int. J. Mental Health. Addict., 1–7.
29. Hjálmsdóttir, A., Bjarnadóttir, V.S.: "I have turned into a foreman here at home": Families and work–life balance in times of COVID-19 in a gender equality paradise. Gend. Work. Organ. **28**(1), 268–283 (2021)
30. Wolor, C.W., Solikhah, S., Fidhyallah, N.F., Lestari, D.P.: Effectiveness of E-training, E-leadership, and work life balance on employee performance during COVID-19. J. Asian Fin. Econ. Bus. **7**(10), 443–450 (2020)
31. Sethi, B. A., Sethi, A., Ali, S., & Aamir, H. S. (2020). Impact of Coronavirus disease (COVID-19) pandemic on health professionals. Pakistan J. Med. Sci. **36**(COVID19-S4), S6.
32. Majumder, S., & Biswas, D. (2021). COVID-19: impact on quality of work life in real estate sector. Qual. Quan., 1–15.
33. Irawanto, D.W., Novianti, K.R., Roz, K.: Work from home: measuring satisfaction between work-life balance and work stress during the COVID-19 pandemic in Indonesia. Economies **9**(3), 96 (2021)
34. İlkkaracan, I., Memiş, E.: Transformations in the gender gaps in paid and unpaid work during the COVID-19 pandemic: findings from Turkey. Fem. Econ. **27**(1–2), 288–309 (2021)
35. Anwer, M. (2020). Academic labor and the global pandemic: revisiting life-work balance under COVID-19. Susan Bulkeley Butler Center for leadership excellence and advance working paper series, **3**(1), 5–13.
36. Matulevicius, S.A., Kho, K.A., Reisch, J., Yin, H.: Academic medicine faculty perceptions of work-life balance before and since the COVID-19 pandemic. JAMA Netw. Open **4**(6), e2113539 (2021)
37. Dewey, C., Hingle, S., Goelz, E., & Linzer, M. (2020). Supporting clinicians during the COVID-19 pandemic.
38. Kelker, H., Yoder, K., Musey, P., Harris, M., Johnson, O., Sarmiento, E., & Welch, J. L. (2020). Longitudinal prospective study of emergency medicine provider wellness across ten academic and community hospitals during the initial surge of the COVID-19 pandemic. Research Square.
39. Mazerolle, S.M., Goodman, A.: Fulfillment of work–life balance from the organizational perspective: a case study. J. Athl. Train. **48**(5), 668–677 (2013)
40. Caligiuri, P.M., De Cieri, H.: Predictors of employeesâ€™ preference for working from home post-pandemic. Bus. Econ. Res. **11**(2), 1–19 (2021)
41. McKim, A. J., Sorensen, T. J., & Burrows, M. (2021). The COVID-19 pandemic and agricultural education: an exploration of challenges faced by teachers. Nat. Sci. Educ. **50**(1), e20060.
42. Chanana, N. (2020). Employee engagement practices during COVID-19 lockdown. J. Public Aff., e2508.
43. Reinwald, M., Zimmermann, S., Kunze, F.: Working in the eye of the pandemic: local COVID-19 infections and daily employee engagement. Front. Psychol. **12**, 2544 (2021)
44. Dwiedienawati, D., Tjahjana, D., Faisal, M., Gandasari, D., & Abdinagoro, S.B. (2020). Transformational leadership, communication quality influences to perceived organization effectiveness and employee engagement and employee retention during the covid-19 pandemic. J. Adv. Res. Dyn. Control Syst. **12**(7 Special Issue).
45. Nguyen, H.N., Tran, M.D.: The effect of perceived organizational support on employee engagement during the COVID-19 pandemic: an empirical study in Vietnam. J. Asian Fin. Econ. Bus. **8**(6), 415–426 (2021)
46. De-la-Calle-Durán, M.C., Rodríguez-Sánchez, J.L.: Employee engagement and wellbeing in times of COVID-19: a proposal of the 5Cs model. Int. J. Environ. Res. Public Health **18**(10), 5470 (2021)

47. Elizur, D., Shye, S.: Quality of work life and its relation to quality of life. Appl. Psychol. **39**(3), 275–291 (1990)
48. Smith, D., Choudhury, P., Chen, G., & Agarwal, R. (2021). Weathering the COVID storm: thes effect of employee engagement on firm performance during the COVID pandemic. Available at SSRN 3841779.

Chapter 2
Data Analysis and Mathematical Modeling for COVID-19 Pandemic

Abstract Data analysis is nothing more than the ability to collect and organize data and translate into meaningful information. Data analysis is a simple form of skill, referring to the use of logical and analytical reasoning to evaluate collected data. Data analysis and mathematical modeling help us to formulate ideas and identify underlying assumptions. The following chapter gives an overview about the need of data analysis and mathematical modeling to understand the impact of the pandemic.

2.1 Introduction

Mathematics plays a significant role in formulating and communicating complex knowledge and hypotheses. Mathematical models are extensively used in natural sciences, engineering disciplines and social sciences. Physicists, computer scientists and economists frequently use mathematics for solving real-life problems. Dynamical systems, statistical models, differential equations, game theoretic models, etc., are various forms of mathematical models. Decision variables, input variables, state variables, exogenous variables, random variables and output variables are basic types of variables used in mathematical models. Variables are generally represented by vectors as there can be many variables of each type. Based on a priori information available for the system, mathematical modeling problems are often classified into two types, namely black box or white box. No priori information is available for black box model, and all necessary priori information are available for a white box model (aka. glass box/clear box). This concept only works as an intuitive guide for approach as all systems are somewhere between the black box and white box models in reality. To make the model more accurate, it is always preferable to use priori information as much as possible for the same.

Data analysts use different statistical models (is a mathematical representation) so as to comprehend and analyze the information from the data effectively [1]. Rather than searching through raw data, this strategy allows data analysts to spot correlations between variables, make predictions about future sets of data and display that data in a way that non-analysts and stakeholders can consume and comprehend it.

G. R. Shinde et al., *Quality of Work-Life During Pandemic*, Studies in Big Data 100,
https://doi.org/10.1007/978-981-16-7523-2_2

Statistical models come in a variety of categories, and a skilled data analyst should be familiar with all of them. Data analysts need to determine not only the best fit model to answer the topic at hand but also the best suitable model to the data to deal with each circumstance [2]. Mathematical modeling can be utilized for a variety of purposes. The state of knowledge about a system, as well as how well the modeling is done, determines how successfully any given purpose is attained. The following are some examples of the types of goals that can be pursued:

1. Developing scientific understanding—by quantitatively expressing present system knowledge;
2. Test the effect of changes in a system;
3. Aid decision making ((i) tactical decisions by managers, (ii) strategic decisions by planners).

In this study, collected data from different sectors (e.g., education, real estate, agriculture, etc.) is analyzed to understand the impact of the pandemic. Data analysis and mathematical modeling help us to formulate ideas and identify underlying assumptions.

2.2 Need of Data Analysis and Mathematical Modeling

Data analysis is nothing but the ability to collect as well as to organize the data and also translate it into meaningful information. Data analysis is the skill in the simple form which refers to using logical and analytical reasoning in order to evaluate the collected data. In the organization when they have to collect data, at that time they monitor its competitors' performance, sales figures and buying trends, etc., by using more strength and effort to be more economical. So many times, no one can understand the customer's behavior as well as competitor's performance without any skills to analyze all the data. That is the reason that data analysis is necessary for making well-informed and efficient decisions. Also, data analysis helps many organizations to define their positions in the market relative to competitors and that helps us to avoid risks as well as it increases opportunities that must be grabbed in order to grow. The main use of data analysis is for the company to come up with new products and services that provide greater satisfaction to them and that allows us to gauge the satisfaction level of the customers and come up with new products. Hence, data analysis is important for the success of businesses [3].

Mathematics is the only one language of science and technology. These days mathematics and mathematical modeling (MM) have been in limelight due to its wide applications to resolve technical as well as non-technical. We can express any situation or real-world problems into mathematical models or equations by using the thousands of known mathematical results, theorems, symbols. Mathematical modeling helps to study an effect of change of a variable or parameter from the set of equations. Analogically, mathematical modeling is an art of winning war without actually fighting. MM is the technique to see, study and understand which is invisible to non-mathematical people.

During pandemic, mathematical modeling is used to find growth rate, recovery rate, mortality rate, etc. Moreover, MM is used for the prediction of the growth rate, necessity of the desired infrastructure, economical change and many more.

To study and understand how the world works and how to make it better, we need mathematical modeling. With the proper study, MM helps to save lives, assists to make better policies and decision making. Additionally, MM can also be used to understand the universe and the conditions needed to sustain life.

In MM, we translate our beliefs into the language of mathematics.

- Mathematics is a very precise language. This helps us to formulate ideas and identify underlying assumptions.
- Mathematics is a concise language, with well-defined rules for operations.
- All the results that mathematicians have proved over hundreds of years are known to us.
- Computers can be used to perform numerical calculations.

Consider the following problems:

- Find the height of a mountain or tower (without climbing it!).
- Find the width of a river or lake (without crossing it).
- How to find the mass of the Earth (without using a balance).
- How to find the temperature of the Sun at its center of surface.
- Find the volume of blood inside the body of a person (without bleeding him to death).
- Predict the population of the world in the year 2050.
- Predict the total amount of insurance claims a health company has to pay in the year 2025.
- To measure severity of the pandemic [2].

All such types of problems have been solving through mathematical modeling only.

2.3 Tools for Mathematical Modeling

Depending upon the nature of the real-world problem, there are following tools for mathematical modeling.

- Mathematical models can be classified according to the subject matter of the models. That is, mathematical modeling in health science, economics, computer engineering, physics, etc.
- MM according to used mathematical technique.
- Mathematical model according to purpose, that is, MM for prediction, optimization, description…

- MM as per the linear or nonlinear nature.
- MM according to static or dynamics.
- MM according to discrete or continuous.
- MM according to deterministic or stochastic.

The following mathematical tools are used for mathematical modeling of real-world problems.

- Set Theory: Classical set theory, rough sets, soft sets, etc., are used for mathematical modeling
- Probability theory
- Graph theory

2.4 Pragmatic Approach for Mathematical Modeling

For greater understanding of the mathematical models existing in the subsequent chapters, the following mathematical theory is required.

2.4.1 Rough Set Theory

Rough set theory is a mathematical tool which deals with imperfect knowledge. The problems associated with imperfect knowledge have been resolved for a long time by philosophers, logicians and mathematicians. There are many mathematical theories which are useful for handling vagueness or imperfect knowledge. Z. Pawlak defined an innovative approach of rough set theory to study, understand and interpret imperfect information.

By using topological operations, interior and exterior approximations, and boundary, the concept of rough set is defined. For the better understanding of the rough set theory, need to learn relations first [4].

A relation on set H is any subset of $H \times H$; i.e., Cartesian product of H with H. A relation R on set H is said to be reflexive if hRh is in H for all h in H. R is said to be symmetric if hRk is in H implies kRh is also in H for all h, k in H. For any h, k, p are in H, and hRk, kRp then hRp, such relation R is called transitive relation on H.

A relation R on set H which is reflexive, symmetric and transitive is called an equivalence relation.

Let $H = \{$set of all straight lines in a plane$\}$ and $R = \{hRk/a$ straight line h is parallel to $k\}$.

This relation R is reflexive, symmetric and transitive. Thus, R is an equivalence relation on H.

Let $L = \{hRk/a$ straight line h is perpendicular to $k\}$ is a relation on H. The relation R is symmetric but not reflexive, not transitive. Hence, R is not an equivalence relation.

Let U be the universal set and H be a non-empty finite set of an attributes. Therefore, $S = (U, H)$ is an information system. Let $P \subseteq H$ and h is in H. An Indiscernibility relation is defined as $\mathbf{Ind_s(P)} = \{(u, v)/h(u) = h(v)$, for all h in $H\}$.

A set of objects which are certainly in Y is called a R-lower approximation of Y with respect to R. A Relation R_* (Y) is called the R-lower approximation of set Y with respect to R if

$$R_*(Y) = \{y/R(y) \subseteq Y\}.$$

A set of objects which are possibly in Y, is called a R-upper approximation of Y with respect to R. It is called R-upper approximation if $R^*(Y) = \{y/R(y) \cap Y \neq \emptyset\}$.

A set of objects which is in R^* (Y) but not in R_* (Y) is called a boundary set of Y with respect to R. That is, A boundary region is given by $B_R(Y) = R^*(Y) - R_*(Y)$.

A set Y is said to be rough set with respect to R if its boundary set is non-empty, i.e., $B_R(Y) \neq \emptyset$.

A set Y is said to be crisp set with respect to relation R if its boundary region is empty, i.e., $B_R(Y) = \emptyset$.

The pictorial representation of the rough set is given in Fig. 2.1. A rough set is also defined by using rough membership function μ_Y^R: $U \rightarrow [0, 1]$

$\mu_Y^R(y) = \frac{|Y \cup R(Y)|}{|R(Y)|} \ldots$, where $|R(Y)|$ denotes the cardinality of $R(Y)$.

The value of the membership function is known as the degree of certainty to which y belongs to Y. It is a kind of the conditional probability.

The rough membership function is depicted in Fig. 2.2.

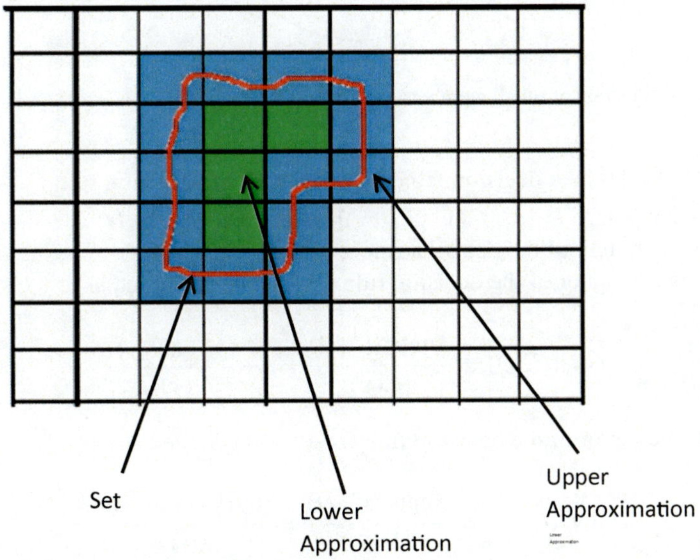

Fig. 2.1 Diagrammatic representation of rough set

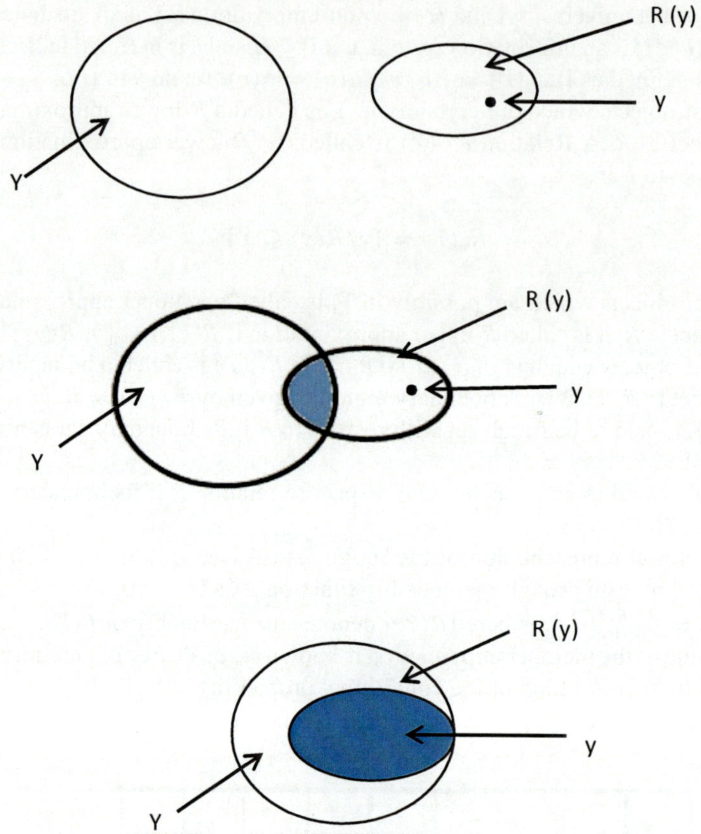

Fig. 2.2 Representation of rough membership function

Let S (U, C, D) be a decision table. A support of the decision rule $C \rightarrow_y D$ is defined as $\sup p_y(C, D) = |C(y) \cup D(y)|$. A number $\sigma_y(C, D) = \frac{\text{supp}_y(C, D)}{|U|}$ is called as the strength of the decision rule $C \rightarrow_y D$.

The certainty factor of the decision rule $C \rightarrow_y D$ is defined as

$$\text{cer}_y(C, D) = \frac{\text{supp}_y(C, D)}{|C(y)|} = \frac{|C(y) \cap D(y)|}{|C(y)|}$$

A coverage factor of the decision rule $C \rightarrow_y D$ is defined as

$$\text{cov}_y(C, D) = \frac{\text{supp}_y(C, D)}{|D(y)|} = \frac{|C(y) \cap D(y)|}{|D(y)|}$$

The readers are advised to understand rough set theory in detailed please refer [3].

2.4.2 Conditional Probability

Probability plays important role in the development of the mathematics as well as statistics. To find the probability of an event, we can apply simple definition of the probability. But, to find the probability of an event with some additional conditions need different concepts of the probability. Such types of the questions are resolved by using a particular concept of the probability, known as the conditional Probability.

Definition: Let S be a sample space. Let H and K be any events of the sample space S. The conditional probability of H given an event K is defined as $P(H/K) = \frac{P(H \cap K)}{P(K)}$, provided $P(K) \neq 0$. Therefore, $P(H \cap K) = P(H/K)P(K)$.

Bayes Theorem: If $H_1, H_2, H_3, \ldots, H_n$ are mutually exclusive events with $P(H_i) \neq 0$ for $i = 1, 2, 3, \ldots, n$ of a sample space S. Let K be any arbitrary event of S with $P(K) > 0$, then

$$P(H_i/K) = \frac{P(H_i)P(K/H_i)}{\sum_{i=1}^{n} P(K/H_i)P(H_i)}$$

2.4.3 Graph Theory

The development of graph theory is similar to the development of probability theory. The graph theory because of its application in many disciplines has emerged as one of the useful branch of mathematics during the last decade. Moreover, any type of virus graphs (COVID-19) is expressed in terms of graphs.

A graph is an ordered pair $(V(G), E(G))$ where

(i) $V(G)$ is non-empty finite set of elements known as vertices or nodes. $V(G)$ is called the vertex set.
(ii) $E(G)$ is a family of unordered pairs (not necessarily distinct) of elements of V, known as edges or arc or branches of G. $E(G)$ is known as edge set.

A graph having finite number vertices and edges is called the finite graph. A graph H which is free from cycles is known as acyclic graph [5].

A graph G (V, E) in which some weight is assigned to every edge of G is called **weighted graph**.

Let G be the graph having vertex set as the set of district places in any state, and edge set is the set of routes joining any two districts. Weight of an edge is the distance between the end points or districts. Hence, the graph G is the weighted graph.

If we consider the directions of the routes, whether routes are one way or two ways, such a graph is known as the directed graph.

A directed, finite and acyclic graph F (V, E, Φ) is called a flow graph if V is the set of vertices, $E \subseteq V \times V$ is a set of directed edges and $\Phi : E \rightarrow_y R^+$ is a flow function, R^+ is the set of non-negative real numbers. If an edge $(h, k) \in E$, then h is an input of k and k is an output of h.

2.5 Summary

Data analysis is the skill in the simple form which refers to using logical and analytical reasoning in order to evaluate the collected data. Data analysis is necessary for making well-informed and efficient decisions. Also, data analysis helps many organizations to define their positions in the market relative to competitors. The main use of data analysis is for the company to come up with new products and services that provide greater satisfaction to them. Hence, data analysis is important for the success of businesses. Mathematics is a very precise language, and this helps us to formulate ideas and identify underlying assumptions. We can express any situation or real-world problems into mathematical models or equations by using the thousands of known mathematical results, theorems, symbols. Mathematical modeling helps to study an effect of change of a variable or parameter from the set of equations. In this chapter, the need of data analysis and mathematical modeling is presented. Various tools used for mathematical modeling are also discussed in this chapter. Data analysis and mathematical modeling help us to formulate ideas and identify underlying assumptions.

References

1. https://www.sciencedaily.com/terms/mathematical_model.htm. Accessed 17 Aug 2021
2. Bhapkar, H.R., Mahalle, P.N., Dey, N., Santosh, K.C.: Revisited COVID-19 mortality and recovery rates: are we missing recovery time period? J. Med. Syst. **44**(12), 202 (2020). https://doi.org/10.1007/s10916-020-01668-6
3. Kapur, J.N.: Mathematical Modelling. Wiley, New York, NY (1988)
4. Deja, R.: Conflict analysis, rough set methods and applications. In: Polkowski, L., Tsumoto, S., Lin, T.Y. (eds.) Studies in Fuzzyness and Soft Computing, pp. 491–520. Physica-Verlag, A Springer-Verlag Company (2000)
5. Harary, F.: Graph Theory. Addison-Wesley, Reading, MA (1969)

Chapter 3
Influence of COVID-19 on Quality of Work-Life in Real Estate Sector

Abstract The real estate sector helps to create high profits in the country's national income, thereby strengthening the economy. The pandemic not only affected the workforce, but also affected the stakeholders of the real estate sector, namely raw material suppliers, builders, building owners, people who participated in promotional activities, etc. Therefore, feelings, expressions, attitudes of real estate professionals are very important factors in work-life. The quality of life for real estate professionals has been hampered. This chapter is an attempt to understand the influence of COVID-19 on quality of work-life in real estate sector using data analysis and mathematical modeling.

3.1 Introduction

Real estate sector is one of the booming sectors in the country. It helps to create high margin of country's national income and thus makes the economy strong. This industry has been widely recognized across the globe. It has three segments, namely residential real estate, commercial real estate and industrial real estate. Residential includes housing, commercial includes office spaces, retails, shopping malls, hospitals and industrial includes factories, plants, warehouses, etc. [1]. The growth of this sector is related to the growth of the business environment. It is estimated that by 2025, the real estate sector will also account for 13% of the country's GDP. Research shows that India's real estate industry will increase to a compound annual growth rate of 19.5% in 2028. The real estate market is expected to grow to Rs. 65,000 crores by 2040 [2]. Under the construction industry, real estate sector is not exceptional to get impacted of the COVID-19 pandemic. The labor market, business liquidity, delivery of project, supply of raw-materials, other key cost components, etc., faced a harmful effect due to the worst effect of pandemic. In the area of completion of projects, scheduling and financing this sector has met many more complexities; i.e., labors were not available, and material was not accessible. Due to non-availability of sufficient numbers of staffs and skilled laborers the planned work could not run smoothly [3]. Lockdown is implemented by various states in India to slow down infection rate of COVID-19. Though infection rate is slow down due to lockdown,

G. R. Shinde et al., *Quality of Work-Life During Pandemic*, Studies in Big Data 100,
https://doi.org/10.1007/978-981-16-7523-2_3

however, it made difficulties for real estate sector; as this sector is labor intensive, large numbers of skilled, semi-skilled, unskilled workers are involved in this field [4]. Stay home strategy was applied every state; hence, many labors migrated toward their hometown; this made unavailability of workers for construction, transport, etc. The international borders were also sealed that made the lack of material required for constructions. The government offices were closed that also made impact as the legal permission documents were delayed. Overall, there was unavailability of the raw material, human resources, legal permissions and many more. This pandemic impacted not only on labors but to all stakeholders of the real estate sector, i.e., raw material suppliers, architectures, construction owners, people involved in promotional activities, etc. Therefore, the feeling, expression, attitudes, expectations of the real estate professionals toward the work-life are very prime factor. Apart from that the quality life of real estate professionals has been hampered. It is already established that human resources are the lifeline of any business organization. These assets need to be taken care of by the employers all the time not only the crisis period. In spite of having physical, technological and financial resources, the need of human resources is highly impacted on the organizational efficiency. By motivating and providing job satisfaction to them, real estate sector can achieve its objectives in terms of image, recognition and revenue generation [5–7]. The quality of work-life is very striking part of real estate professionals at this pandemic time. This gives a positive attitude to the working class. It helps to increase productivity and improves the standard of living [8]. The quality of work-life is related to employees' performance, and it has a direct relation with organizational performance.

What is the status of quality of work-life of these employees in these pandemic situations, what difficulties and consequences they face in their professional life and what can be the consequences of it, to find out these phenomena, we mainly focused and collected the data from different firms under real estate sector.

The key contributions presented in the scope of this chapter are listed as follows:

1. The preparation of a questionnaire for employees in real estate sectors for recording their ongoing experiences during this COVID-19 pandemic.
2. Preparation of the dataset based the responses collected from the survey initiated for employees in the real estate sectors.
3. Design a mathematical model to draw statistical inferences.
4. Machine learning-based analysis and visualization to draw useful inferences and normative interpretations from the dataset.

Section 3.2 reports literature survey of this study. The description of data and dataset is discussed in Sect. 3.3, and statistical description of the dataset is presented in Sect. 3.4. Mathematical modeling is presented in Sect. 3.5. Under Sect. 3.6, data analysis and visualization is reported. Section 3.7 is the summary of the overall study.

3.2 Literature Survey

The social distancing and lockdown made it difficult for real estate sector, and various constructors have faced economical loss. Few constructors may have availability of labors but not having access to raw material, few are having raw material but facing lack of human resources, and in all cases, the construction work has been stopped for long period. The researchers have started looking to understand the impact of pandemic on real estate and trying to find the mitigation techniques that can improve the situation. Tanrıvermiş presented the impact of this pandemic of the real estate sector in Turkey [9]. The study presented the impact on GDP and housing prices. The analysis shows that there will be significant change in the way of investments, and people are interested in where they will get fixed returns for their investment, i.e., fixed deposits, gold, foreign exchange, etc. Uchehara et al. [10] presented the impact of COVID-19 and risk management for the same. In this study, the mitigation techniques for tenant and house owners are given, and in pandemic, there was huge effect on rent of houses. Various rules were applied in the favor of tenants in the period of lockdown. However, there should be rules for negotiation in such situations where both parties can come down on some agreement. Giudice et al. [11] presented the impact of pandemic on real estate sector of Italy. The study presents that the demand of housing price as well as requirements will change significantly, the home isolation, social distancing and work from home strategies have focused on need of more space in home. This will impact of housing demands and prices. Milcheva [12] presented the impact of pandemic on USA and Asia. Study shows that the real estate sector impacted varied in USA and Asia. In initial few months of pandemic, US real estate impacted hugely than Asia, and less negative impact is observed in Asian countries. There is also difference in fields where impact has been observed, and the retail and hotel were impacted more in USA, and in Asia, offices were impacted. Marona and Tomal [13] have presented the impact of pandemic on house brokers. Social distancing made broker's to shift digital platform than from their traditional way of working. The study shows that there is significant change in requirements of landlord and tenants, and it is also observed that tenants are asking high-end facilities with low rent. In recent time, a lot of work has been carried out to understand the impact of this pandemic on the real estate sector and economy [14–19]. In this pandemic time, the quality of work-life of employees has become one of the significant issues. Singh et al. [20] focused in their study on discovering the physical and mental exhaustion and difficulties faced by the employees during pandemic. They made solution for it by providing the phenomenon of work-life integration rather than work-life balance. The employees who accepted these changes and made familiar themselves with the digitalization were capable to meet all the demands in relation to home, family, time, work, mental–physical health and responsibilities. The concept working from home has been generated in this pandemic time, and it has penetrated the professional and personal spaces a lot of the employees. The working time becomes longer. More usage of digital platforms makes the work-life harder. Deshpande et al. [21] made the study on work-life integration.

Bhapkar et al. [22] gave mathematical model to reduce symptoms of COVID-19 by using rough set theory approach. Bhapkar [23] presented any types of viruses by using graphs and defined growth rates of viruses in terms of the corona product of graphs. Bhapkar et al. [24] defined the first generic formulae to measure the severity of any epidemic or pandemic during its period also, known as the progressive recovery rate and progressive mortality rate.

Therefore, from the above literature, it can be seen that so many studies have been conducted on construction industry as well as real estate sector. There is a need to find out the status of quality of work-life in real estate sector. We have already mentioned that there is a relation between qualitative work-life and employees performance. This phenomenon is also important to know based on this pandemic situation. Employers will take the necessary actions on the basis of these findings. To keep in mind, we have focused on influence of COVID-19 on quality of work-life in real estate sector.

3.3 Dataset and Data Description

Survey is carried out for employees of middle-tier real estate companies in Kolkata, West Bengal, India. The data were collected during the period, May 2020 to July 2020, and 107 employees from the real estate sector in varied profiles have responded to the survey.

The questionnaire consists of the total number of 17 questions which consist of 5 questions related to know demographic information and 12 questions are regarding current recruitment policy, employee morale, online learning process that has been arranged by the company, their work performance during this time, the status of mental health, company's safety norms, any insurance policy on COVID-19 for affected employees, the status of bonus, pay cut, career growth opportunities, termination and organization's commitment. The questionnaire was distributed through e-mail among jobholders in the real estate sector. The questionnaire is shown in Table 3.1.

3.4 Descriptive Statistics

Dataset can be described briefly with the statistical information. Statistical information includes measure of central tendency and measure of data dispersion. The standard deviation is the measure of data dispersion and the mean, mode, median are the measure of the central tendency of the dataset. Based on the mean, median and mode values symmetry of data can be understood. The central tendency of the dataset describes how data is distributed? How it is spread from its mean value? What is the answer that maximum times appeared in responses of participants?

The qualitative statistical description of this dataset is shown in Table 3.2. The Pandas library in Python is used to calculate statistical values of each.

Table 3.1 Questionnaire

Sr. No	Questions
1	Age
2	Qualification
3	Managerial position
4	Experience
5	Gender
6	Does your company use Artificial Intelligence (AI)-driven technology to manage recruitment function in this new normal?
7	Does the management maintain the accountability or employee morale to keep it up in this pandemic scenario?
8	Do you agree employees are encouraged to up skilling themselves with online learning platform in this pandemic time?
9	Is the company taking care of your wellness including mental health in this present time?
10	Does your company use new method or technology to manage employee's performance in this present time?
11	Do you agree that your organization is following strict safety measures (as per WHO/World Health Organization) guidelines to protect the workers from the pandemic?
12	Do you get any insurance coverage from the protection of COVID-19 disease by your organization?
13	Does your organization provide performance related pay or bonus in this present scenario?
14	Do you face pay cut in this pandemic time?
15	Is your organization giving emphasize on career planning or promotional activities in this situation?
16	Does your company follow the decision of termination and retrenchment in this time?
17	Do you agree your organization has commitment to you in this pandemic time?

Table 3.2 Statistical values

Demographic/variables	Standard deviation	Median	Mean	Min value	Max value
D_1	0.678287004	2	2.03809524	1	4
D_2	0.83808171	2	1.76190476	1	4
D_3	0.58475755	2	2.18095238	1	3
D_4	0.622090296	2	2.3047619	1	3
D_5	0	1	1	1	1
V_1	1.031375908	2	2.11428571	1	4
V_2	1.046273695	4	3.43809524	1	5
V_3	0.790858948	4	3.9047619	2	5
V_4	1.193925406	4	3.63809524	1	5
V_5	0.990152983	2	2.78095238	1	5

(continued)

Table 3.2 (continued)

Demographic/variables	Standard deviation	Median	Mean	Min value	Max value
V_6	1.051773315	4	3.9047619	1	5
V_7	1.282239913	2	2.32380952	1	5
V_8	1.468347355	4	3.51428571	1	5
V_9	1.3227372	4	4.01904762	1	5
V_{10}	1.179805659	3	2.61904762	1	5
V_{11}	1.246386719	3	3.15238095	1	5
V_{12}	1.081140356	3	2.81904762	1	4

3.5 Mathematical Modeling

In this section, we make mathematical model for the data of the following questions.

Que. 3. Managerial Position of the Respondents.

(1) Supervisor Level (L)
(2) Middle Level (M)
(3) Top Level (T)

And

Que. 9. Is the company taking care of your wellness including mental health in this present time?

(1) Strongly Disagree
(2) Disagree
(3) Neither Agree nor Disagree
(4) Agree
(5) Strongly Agree

For the data of Que. 3 and 9 from industrial sector, make the following Table 3.3 with support, strength factor, certainty factor and the coverage factor.

From the certainty factor, we can make the following conclusions.

Table 3.3 Strength, certainty and coverage factors

Managerial position	Mental health	Support	Strength factor	Certainty factor	Coverage factor
General/supervisor level	1	0	0.000	0.000	0.000
	2	0	0.000	0.000	0.000
	3	2	0.019	0.069	0.133
	4	27	0.257	**0.931**	**0.643**
	5	0	0.000	0.000	0.000

(continued)

Table 3.3 (continued)

Managerial position	Mental health	Support	Strength factor	Certainty factor	Coverage factor
Middle level	1	0	0.000	0.000	0.000
	2	14	0.133	0.212	**1.000**
	3	13	0.124	0.197	**0.867**
	4	12	0.114	0.182	0.286
	5	27	0.257	**0.409**	**1.000**
Top level	1	7	0.067	**0.700**	**1.000**
	2	0	0.000	0.000	0.000
	3	0	0.000	0.000	0.000
	4	3	0.029	0.300	0.071
	5	0	0.000	0.000	0.000

- 93.1% respondent of general/supervisor level disagree that the company is taking care of their wellness including mental health during this pandemic.
- 40.9% respondent from middle level category strongly disagree that the company is taking care of wellness including mental health during this pandemic.
- 70% respondent from top-level category strongly agree that the company is taking care of their wellness including mental health during this pandemic.
- There is no respondent from general/supervisor level of the opinion that that the company has not taken any care of their wellness including mental health during this pandemic.

The following conclusions are made by using the coverage factor of the data.

- 100% agree responses came from middle level.
- 100% strongly disagree came from middle level.
- 100% strongly agree responses came from top level.
- 64.3% disagree responses came from general level.
- 86.7% neutral responses came middle level.
- 0% agree responses came from top level.

The flow graph associated with Table 3.3 is given below.

The branches of the flow graph represent decision rules together with their certainty factor and coverage factor. For example, the decision rule T ⟶ (1) has the strength factor, certainty factor, and coverage factor 0.067, 0.70 and 1.00, respectively.

The flow graph gives a clear insight into the work-life structure of all the respondents with respect to their positions.

In some applications, the exact values of the certainty and coverage factors of decision rules are very important. The approximate decision rule is denoted by H ⟶ B is defined as follows.

H ⟶ B if and only if certainty factor of H and B is more than 0.5, that is cer(H, B) > 0.5.

By using the approximation rules, we can replace the flow graph as shown in Fig. 3.1 by approximate flow graph presented in Fig. 3.2.

From Fig. 3.2, it is observed that the top-level category respondents mostly approach strongly agree. Moreover, the low-level respondents mostly approach disagree level. The respondents from middle level mostly do not approach to any particular level.

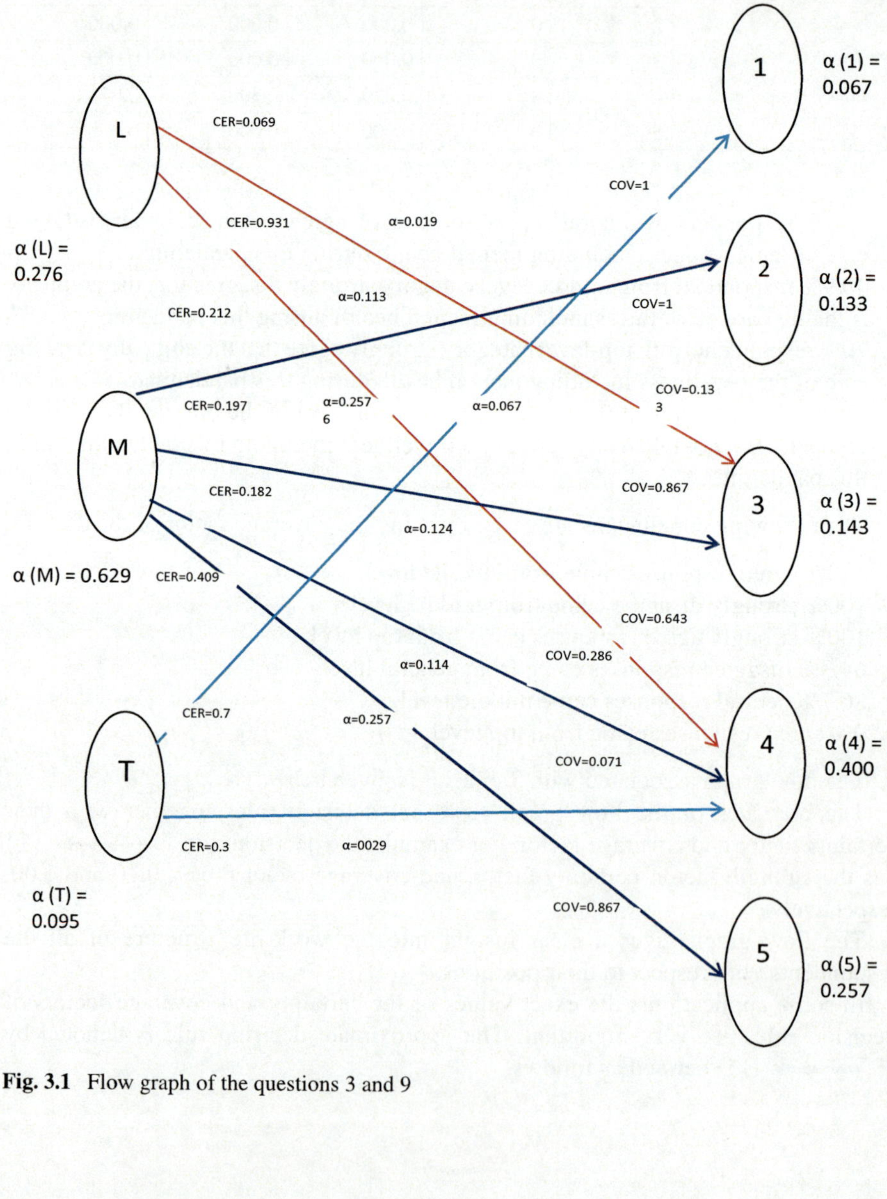

Fig. 3.1 Flow graph of the questions 3 and 9

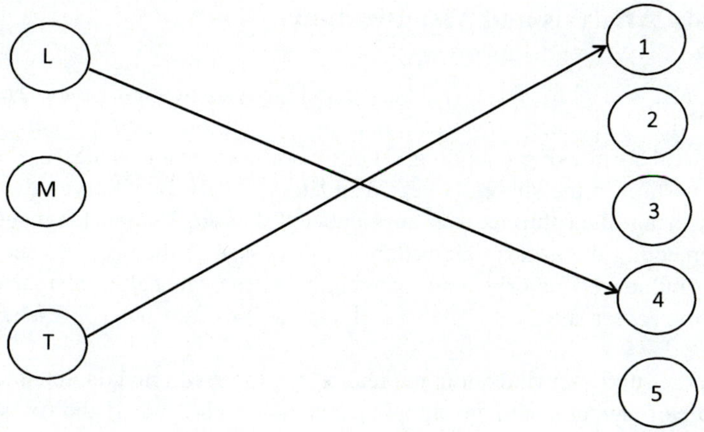

Fig. 3.2 Approximate flow graph

Additionally, from the flow graph shown in Fig. 3.2, we can get an inverse approximate flow graph which is given below in Fig. 3.3.

The above flow graph contains all inverse decision rules with coverage factor greater than 0.5. From the figure, we can conclude that disagree scales were attained mostly from supervisor-level respondents. Agree scales were obtained mostly from top-level respondents.

By considering all the above points, we reached to an important conclusion that during pandemic period, scales of the managerial positions and the mental health support are in the inverse proportion.

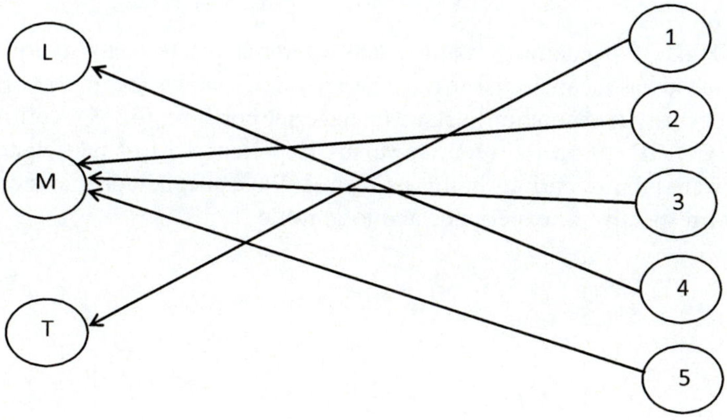

Fig. 3.3 Inverse approximate flow graph

3.6 Data Analysis and Visualization

ML approach is used for the visualization and analysis of survey data. The dataset is in the form excel sheet; however, it is in the textual form, and it is converted into the numerical form using encoding. There were no missing values in the dataset as each question in the survey is kept mandatory. Analysis of survey data is done using a k-mean algorithm as it is an unlabeled dataset; Python language is used for implementing the k-mean algorithm. As discussed in the previous section, few factors are influenced majorly in this pandemic, i.e., mental health, salary deductions, job security, career shift, etc. Analysis of such a few questions is presented in this section.

As per the survey carried out in the real sector, there was no concrete mechanism to take care of mental health during this pandemic. Figure 3.4 (i) and (iv) shows the graphs of mental health impact versus managerial position and experience, respectively. In the survey, mixed reactions are received, and particularly, experienced group of employees has stated that the present mechanism was not adequate for the same. In the view of this, it is recommended that there should be some policies in the place in order to insure good mental health of employees, and in turn, it will also improve efficiency and productivity.

In the COVID-19 pandemic, a significant impact has been seen in the financial domain, businesses were affected, and each organization's economic condition is hampered. Given this, the question regarding financial impact is formed, and an analysis of participants' responses is done.

Figure 3.4 (ii) shows the impact of COVID-19 on salaries versus managerial position Fig. 3.4 (v) shows the impact of COVID-19 on salaries versus experience. Most participants stated that there were salary deductions in this pandemic. Concisely, we can conclude that there is a negative impact on salaries irrespective of experience and job profile.

The COVID-19 pandemic has adverse affected real estate sector majorly; hence, unemployment was the major issue raised; Fig. 3.4 (iii) shows the impact of COVID-19 on job security (termination) versus managerial position; Fig. 3.4 (vi) shows the impact of COVID-19 on job security versus Experience. Most participants stated that there were job insecurities in this pandemic. We can conclude that there is a job insecurity irrespective of experience and job profile

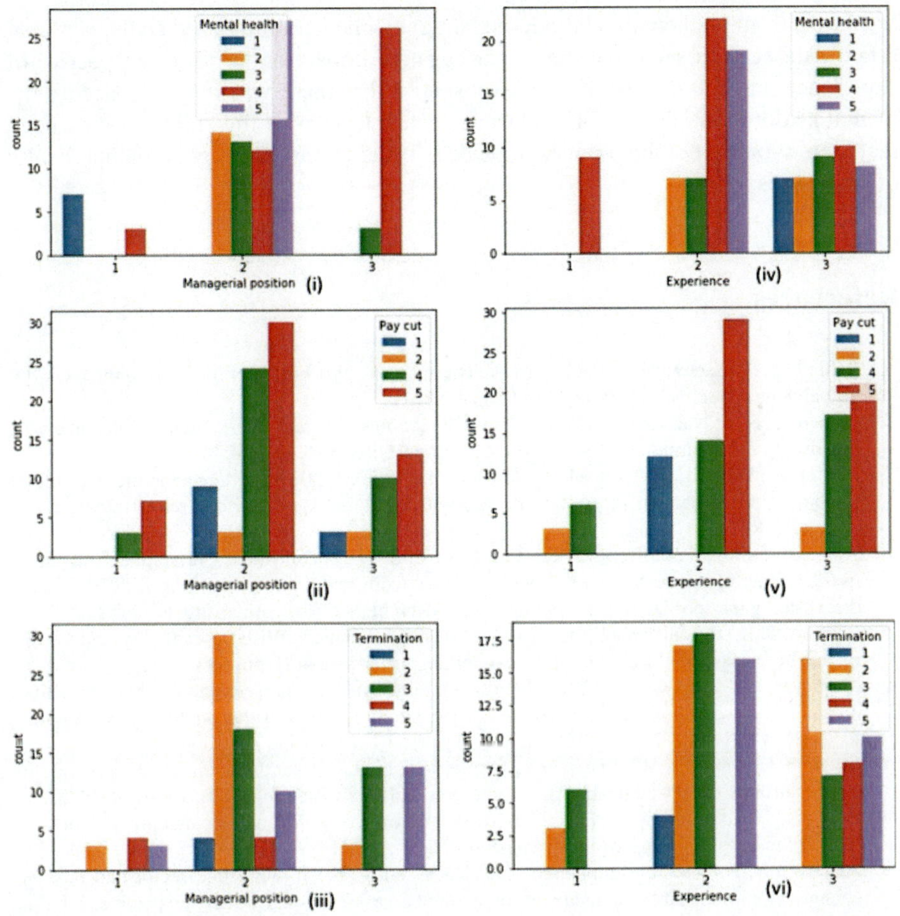

Fig. 3.4 Analysis of impact of COVID-19 pandemic on mental health, pay cut and termination

3.7 Summary

In the COVID-19 pandemic, a significant impact has been seen in the financial domain, businesses were affected, and each organization's economic condition is hampered. Lockdown is implemented by various states in India to slow down infection rate of COVID-19. Though infection rate is slow down due to lockdown however it made difficulties for real estate sector as this sector is labor intensive. The international borders were also sealed that made the lack of material required for constructions. Overall, there was unavailability of the raw material, human resources, legal permissions and many more. The main contributions presented in this chapter are mathematical modeling and ML examination of the data collected from the survey. The visual representation of the analysis will be helpful to the readers for better understanding as well as decision making and recommendations. The conclusions are drawn from mathematical modeling and machine learning-based analysis of real estate dataset. It is seen that there is a negative impact on salaries irrespective of experience and job profile. It is also observed that there was more impact on the mental health of employees of real estate sector. In view of this, it is recommended that there should be some policies in place in order to ensure the good mental health of employees.

References

1. Schieman, S., Badawy, P.J., Milkie, M.A., Bierman, A.: Work-life conflict during the COVID-19 pandemic. Socius **7**, 2378023120982856 (2021)
2. Schieman, S., & Narisada, A. (2021). A less objectionable greed? Work-life conflict and unjust pay during a pandemic. Res. Soc. Strat. Mobility **71**, 100564
3. Martel, A., Day, K., Jackson, M. A., & Kaushik, S. (2021). Beyond the pandemic: the role of the built environment in supporting people with disabilities work life. Archnet-IJAR: Int. J. Architect. Res.
4. Adams-Prassl, Abi and Boneva, Teodora and Golin, Marta and Rauh, Christopher, inequality in the impact of the coronavirus shock: evidence from real time surveys (April 2020). CEPR discussion paper no. DP14665. Available at SSRN: https://ssrn.com/abstract=3594297
5. Putranti, H.R.D., Suparmi, S., Susilo, A.: Work life balance (WLB) complexity and performance of employees during covid-19 pandemic. Arthatama **4**(1), 56–68 (2020)
6. Majumder, S., & Biswas, D. (2021, February). Fundamentals, present and future perspectives of quality of work life. In: Proceeding of First Doctoral Symposium on Natural Computing Research: DSNCR 2020 (vol. 169, p. 271). Springer Nature
7. Mulyani, I., Rahardjo, W., Andriani, I., Qomariyah, N.: Factors influencing employee's quality of life during COVID-19 pandemic. Indigenous. J. Ilmiah Psikologi **5**(2), 164–174 (2020)
8. Rinaldi, M. R., Saifudin, S., & Fahlevi, M. Work from home during the pandemic & work-life balance: married working woman perspective
9. Tanrıvermiş, H.: Possible impacts of COVID-19 outbreak on real estate sector and possible changes to adopt: a situation analysis and general assessment on Turkish perspective. J. Urban Manag. **9**(3), 263–269 (2020)
10. Uchehara, I., Hamma-Adama, M., Obiri, K. A., Jafarifar, N., & Moore, D. (2020). Impacts and risk management of COVID-19 pandemic on real estate supply chain. Int. J. Real Estate Stud. **14**(Special Issue 1)
11. Del Giudice, V., De Paola, P., Del Giudice, F.P.: COVID-19 infects real estate markets: short and mid-run effects on housing prices in Campania region (Italy). Soc. Sci. **9**(7), 114 (2020)
12. Milcheva, S. (2021). Volatility and the cross-section of real estate equity returns during Covid-19. J. Real Estate Fin. Econ., 1–28.
13. Marona, B., Tomal, M.: The COVID-19 pandemic impact upon housing brokers' workflow and their clients' attitude: Real estate market in Krakow. Entrepreneurial Business and Economics Review **8**(4), 221–232 (2020)
14. Ogunnusi, M., Hamma-Adama, M., Salman, H., & Kouider, T. (2020). COVID-19 pandemic: the effects and prospects in the construction industry. Int. J. Real Estate Stud. **14**(Special Issue 2)
15. Hoesli, M., & Malle, R. (2021). Commercial real estate prices and covid-19. Swiss Finance Institute Research Paper, (21–08)
16. Sulaiman, M. Z., Aziz, M. N. A., Bakar, M. H. A., Halili, N. A., & Azuddin, M. A. (2020, December). Matterport: virtual tour as a new marketing approach in real estate business during pandemic COVID-19. In: Proceedings of the International Conference of Innovation in Media and Visual Design (IMDES 2020). Atlantis Press (pp. 221–226)

17. Nanda, A., Xu, Y., Zhang, F.: How would the COVID-19 pandemic reshape retail real estate and high streets through acceleration of E-commerce and digitalization? J. Urban Manag. **10**(2), 110–124 (2021)
18. Majumder, S., & Biswas, D. (2021). COVID-19: impact on quality of work life in real estate sector. Qual. Quan., 1–15.
19. Oey, E., & Lim, J. (2021). Challenges and action plans in construction sector owing to COVID-19 pandemic–a case in Indonesia real estates. Int. J. Lean Six Sigma
20. Singh, A.: Work life balance during covid pandemic. Turkish J. Comput. Math. Educ. (TURCOMAT) **12**(4), 516–520 (2021)
21. Deshpande, A., Salunke, P., Joshi, T.: Work life balance in phase of pandemic. Bi-lingual Int. Res. J. **10**(38), 229–240 (2020)
22. Bhapkar, H.R., Mahalle, P.N., Shinde, G.R. and Mahmud, M. , 2021. Rough sets in COVID-19 to predict symptomatic cases. In: K.C. Santosh and A. Joshi, eds., COVID-19: prediction, decision-making, and its impacts. Lecture notes on data engineering and communications technologies (60). Springer, Singapore, pp. 57–68. ISBN 9789811596810
23. Bhapkar, H.R., Mahalle, P.N., Dhotre, P.S. (2020) Virus graph and COVID-19 pandemic: a graph theory approach. In: Hassanien, A.E., Dey, N., Elghamrawy, S. (eds) Big data analytics and artificial intelligence against COVID-19: innovation vision and approach. Studies in big data, vol. 78. Springer, Cham. https://doi.org/10.1007/978-3-030-55258-9_2
24. Bhapkar, H.R., Mahalle, P.N., Dey, N., Santosh, K.C.: Revisited COVID-19 mortality and recovery rates: are we missing recovery time period? J. Med. Syst **44**(12), 202 (2020). https://doi.org/10.1007/s10916-020-01668-6

Chapter 4
Impact of COVID-19 Pandemic on Education Sector

Abstract The education sector is highly affected by the COVID-19 pandemic, because the traditional teaching process requires students and teachers to meet at one place. Due to the closure, various teaching modes and methods have been adjusted, and the evaluation method has also changed from the traditional way. The ebb and flow of this uncertainty in the distribution of COVID-19 led to poor planning of the teaching process (students and teachers), test methods and assessment techniques, and subject work quality. Ultimately, this affected the placement of students in companies and universities. This chapter is an attempt to understand the impact of COVID-19 on education sector (for different stakeholders such as teachers, students, employers and book printers) using data analysis and mathematical modeling.

4.1 Introduction

Since December 2019 COVID-19 pandemic is emerged as a big problem all over the world, as a result of which various strategies were applied by different countries to tackle the infection of COVID-19. From those strategies, lockdown has been the most important and effective strategy, where various restrictions are imposed on the daily routines of personals around the globe. These restrictions are been effective from the perspective of slowing down the infection rate, but at the same time it made day-to-day life very difficult for all. As a side effect of the above measures, new problems like economic imbalance, mental, social, or personal disturbances are raised [1–3]. It has also made noticeable impacts on sectors, such as manufacturing [4–6], real estate [7], agriculture [8], education [9–14], health [15], public safety (police), and industry and so on. In the whole span of this pandemic the quality of working life is affected to a great extent, affecting efficiency, productivity and quality of life. In recent time, a lot of work has been carried out to understand the impact of this pandemic on the education sector [9–14]; however, it is important to understand the views of employees in the educational sector regarding this pandemic. In line with this, the main objective of this chapter is to analyze the data collected from education sector workers to build a mathematical model based on machine learning to study the impact on human life. In this chapter, the impact of COVID-19 pandemic on

G. R. Shinde et al., *Quality of Work-Life During Pandemic*, Studies in Big Data 100,
https://doi.org/10.1007/978-981-16-7523-2_4

education sector is presented and discussed for different stakeholders of education sector who are affected such as teachers, students, employers and book printers. This chapter focuses on the impact of COVID-19 on the quality of working life of teaching and support staff in engineering colleges in India. The education sector suffers the most from the COVID-19 pandemic as the traditional teaching–learning process requires students and teachers to come together in one place. Due to closure, various teaching–learning modes and methods have been adjusted and assessment methods have also been changed from the traditional way. This uncertain ebb and flow in the distribution of COVID-19 led to poor planning of the teaching–learning process (students and teachers), examination methods and assessment techniques, and the quality of the subject work. Ultimately, this has affected students' placements in companies and universities. For this reason, the questionnaire for this survey was designed in consultation with all stakeholders. The questions focus on the support received by the teachers from the employers, difficulties faced in teacher–learning process. The questionnaire was prepared, and survey responses were collected via Google forms. The questionnaire was emailed to teaching and support staff in engineering colleges. The staff with different experiences, qualifications and job profiles participated in the survey. These questions and their responses are then analyzed using mathematics and machine learning tools to draw conclusions.

The main contribution of this study is:

- The preparation of a questionnaire for employees in education sectors for recording their ongoing experiences during this COVID-19 pandemic.
- Preparation of the dataset resulted from the responses collected from the survey initiated for employees in the education sectors.
- Design a mathematical model to draw statistical inferences, correlation in the form of association rule mining and key observations.
- Machine learning-based analysis and visualization to draw useful inferences and normative interpretations from the dataset.

4.2 Literature Survey

Traditional teaching has been avoided due to the COVID-19 pandemic, the online education has its pros and cons. Many researchers have discussed the impact of this pandemic on the education sector. Sumitra Pokhrel and Roshan Chhetri presented comprehensive literature regarding the impact of COVID-19 on the teaching–learning process [16]. The various challenges for online teaching are presented in this work. Mukesh Rawal presented the impact of COVID-19 on the Indian Education system [17]. The focus of this research is to understand the way in which teachers–students are handling online education, the positive and negative impacts of online education. In this work, challenges like availability on the Internet and electricity in a rural area, creating softcopy of learning material, digital literacy are discussed. Magomedov et al. presented the negative and positive impact of a pandemic on education globally [18]. Study focus that there is a difference between online conduction of theory and practical sessions, conduction online learning for theory lectures may not affect that much than practical assignment implementations. Jena has presented the impact of a pandemic on education in India [19]. In this study, the reasons for positive and negative impacts on education due to online educations are discussed, and few suggestions are also provided for the improvement of online education. Joshi et al. presented barriers faced by students and teachers in online education [20]. Teachers have face various challenges like lack of facility to take a lecture from home (separate rooms), lack of training provided by the employer, lack of funding to purchase new technologies, distractions in teaching as they were teaching from home and various other technical difficulties. Ferrel et al. presented the impact of a pandemic on medical education [10], medical education is difficult to carry out online as a student needs to learn by assessing patients physically. In this pandemic due to online education medical students have also missed opportunities to go to conferences as various conferences were canceled due to pandemic. Quezada et al. [16] presented impact of pandemic on teacher of *California liberal arts college, and* study presents challenges faced by teachers in online education. This study also presents few recommendations to improve the online education. Flores et al. [17] presented the impact of pandemic on teaching on institute's of Portugal. The study discusses the challenges while conducting the practice sessions, this study also presents the strategies to improve monitoring of online teaching. Bhapkar et al. [18] given mathematical model to reduce symptoms of COVID-19 by using rough set theory approach. Bhapkar et al. [19] presented any types of viruses by using graphs and defined growth rates of viruses in terms of the corona product of graphs. Bhapkar et al. [20] defined the first generic formulae to measure the severity of any epidemic or pandemic during its period also, known as the progressive recovery rate and progressive mortality rate.

The impact of pandemic on education field is discussed in the literature; however, it is important to know opinion of teachers regarding the online education. In view of this, the survey of teacher's opinions is conducted and analysis is done.

The chapter is organized as follows: The description of data and dataset are discussed in Sect. 4.3, statistical description of the dataset is presented in Sect. 4.4. The mathematical modeling for the educational dataset is designed and presented in Sect. 4.5. The data is analyzed and visualized using ML techniques, presented in Sect. 4.6

4.3 Dataset and Data Description

A survey was conducted in June 2021 and 335 employees from the education sector with different profiles participated. The questionnaire consists of 17 questions; five of them are for demographic information and other questions are related to the impact of COVID-19 on the quality of working life. It is intended that survey participants respond to the questions by selecting one of the answers from five response options. The responses are in the form of fuzzy linguistic variables/values that scale from low–high to make it easier for respondents to complete the survey while having fun with the survey process.

The questions are divided into three categories, namely demographic questions, employee-centered questions and employer-centered questions. The employer-centered questions relate to how the employer has supported its employees in this pandemic situation. The employee-centered questions refer to how the education sector employees have dealt with the pandemic and how it has affected the quality of their work-life; these are presented in Table 4.1.

In summary, the excel sheet of responses, i.e., dataset, consists of 335 samples and 17 features. Among that, 5 are demographic features, and the remaining 12 are variables.

Table 4.1 Questionnaire

Category of questions	Questions
Demographic	Age
	Qualification
	Gender
	Job Profile
	Experience
Employee-centric	**Mental health:** How much COVID-19 pandemic situation impacted your mental health and wellbeing?
	Teaching: To what extent your teaching was satisfactory on this digital platform?
	Impact: COVID-19 pandemic impacted all sectors; how you rate the educational industry in this concern
	Career shift: If this pandemic situation made you think about a career shift, express your thought on the below scale
	Family crisis: Due to the uncertain situation from last year, is family crisis increased and to what extent you can overcome those
	Skill sets improvement: Rate the improvement in your skill sets and knowledge during this pandemic?
	Work-life balance: How much you can make work-life balance in this ongoing situation rate on below scale
	Learning performance: As per your opinion, what are students' learning performance and assessment on this digital platform?
Employer-centric	**Security by company**: Rate additional support provided by your current employer regarding your and your family's security
	Job security: Extent to which COVID-19 pandemic hampered your job security?
	Pay cut: Impact of a pandemic on your salaries from employer, select your satisfaction of scale given below
	Arrangements: Are you satisfied with arrangements made by the employer for the conduction of your duties in the pandemic situation?

4.4 Descriptive Statistics

Statistical information of the dataset helps to briefly describe the dataset, which allows the dataset to be summarized. Statistical information can describe the central tendency of the dataset; i.e., how are the data distributed? How do they scatter from their mean? What is the response given by most participants for a particular characteristic (mode)? What is the mean value of a particular characteristic? The qualitative statistical description of this dataset is presented in Table 4.2. The statistical values of each feature are calculated using the Pandas library in Python; the features are renamed U_1 to U_{17}, and further sections of the chapter will use these naming conventions.

The mean, mode and median are the measures of the central tendency of the dataset, and the standard deviation is the measure of the dispersion of the data. For symmetric data, the mean, median and mode values are equal and lie in the middle. For positively skewed data, the mode value is less than the mean, and for negatively skewed data, the mode value is greater than the mean. From the mean, median and mode values, the symmetry of the data can be inferred. Here for variables, U_1 to U_7, U_9, U_{10}, U_{12}, U_{16} have excellent mean value than mode and U_8, U_{11}, $U_{13–15}$, U_{17} have smaller mean value than mode.

Table 4.2 Statistical values

Demographic/variables	Standard deviation	Median	Mean	Mode	Min value	Max value
U_1: Age	0.829219	2	2.516418	2	1	5
U_2: Qualification	0.564746	3	3.185075	3	1	5
U_3: Gender	0.476492	1	1.346269	1	1	2
U_4: Job profile	1.028011	2	2.408955	2	1	5
U_5: Experience	1.295958	3	3.134328	3	1	5
U_6: Mental health	1.028549	3	3.104478	3	1	5
U_7: Security by company	1.280508	3	2.483582	1	1	5
U_8: Job security	1.253306	3	2.832836	3	1	5
U_9: Teaching	0.919661	3	3.286567	3	1	5
U_{10}: A pay cut	0.996975	2	1.922388	1	1	5
U_{11}: Impact	1.129477	4	3.686567	4	1	5
U_{12}: Arrangements	1.093428	2	2.385075	2	1	5
U_{13}: Career shift	1.434992	2	2.808955	4	1	5
U_{14}: Family crisis	1.287149	2	2.534328	3	1	5
U_{15}: Skill sets improvement	0.928521	4	3.426866	4	1	5
U_{16}: Work-life balance	0.894617	3	3.089552	3	1	5
U_{17}: Learning performance	0.96681	3	2.602985	3	1	5

4.5 Mathematical Modeling

Generic Mathematical Modeling by Bayes' Theorem.

Let U_1, U_2, U_3, ... U_n be n questions or boxes or urns. Let C_i, $i = 1, 2, 3, \ldots m$, be choices or options of the questions according to likers scale, presented in Table 4.3. The set questions are an equiprobable sample space. Therefore, the probability of selecting the question U_i is

$$P(U_i) = 1/n\,,\ \text{for}\ j = 1, 2, 3, \ldots, n$$

Let N be the total number of responses of the experiment.

Let C_{ij} be the total number of responses noted in favor of C_i for the question U_j. The questions and corresponding choices are presented in Table 4.3.

$$\text{where } T_i = \sum_{j=1}^{n} C_{ij};\ \text{for}\ i = 1, 2, \ldots m.$$

Therefore, the conditional probability of C_i given U_j is defined as

$$P(C_i/U_j) = \frac{C_{ij}}{N};\ \text{for all}\ i\ \text{and}\ j$$

$$\text{And} \sum_{i=1}^{m} P(C_i/U_j) = 1,\ \text{for all}\ j$$

Now, the conditional probability of U_j, given C_i, is defined as

$$P(U_j/C_i) = \frac{P(U_j) \times P(C_i/U_j)}{\sum_{j=1}^{n} P(U_j) \times P(C_i/U_j)};\ \text{for all}\ i\ \text{and}\ j$$

And,

Table 4.3 Question and choices

Options\questions	U_1	U_2	U_3	...	U_n	Total
C_1	C_{11}	C_{12}	C_{13}	...	C_{1n}	T_1
C_2	C_{21}	C_{22}	C_{23}	...	C_{2n}	T_2
C_3	C_{31}	C_{32}	C_{33}	...	C_{3n}	T_3
⋮	⋮	⋮	⋮	...	⋮	⋮
C_m	C_{m1}	C_{m2}	C_{m3}	...	C_{mn}	T_m

$$\sum_{j=1}^{n} P(\mathrm{U}_j/\mathrm{C}_i) = 1; \text{ for all } i.$$

Mathematical Modeling for Work-Life in Educational Sector

Let U_1, U_2, U_3, … U_{17} be questions or boxes or urns. Let C_1, C_2, C_3, C_4, C_5 be choices or options of the questions according to a Likert scale of 1 to 5. Therefore, the probability of selecting the question U_i is

$$P(\mathrm{U}_i) = 1/17, \text{ for } j = 1, 2, 3, \ldots, 17$$

The total number of responses or individuals under consideration for the work-life in an educational sector is 335.

Let C_{ij} be the total number of responses noted favoring C_i for Uj's question, presented in Tables 4.4 and 4.5.

Therefore, the conditional probability of C_i given U_j is defined as

$P(\mathrm{C}_i/\mathrm{U}_j) = \frac{C_{ij}}{335}$; for all i and j

Table 4.4 Responses for each choice

Boxes	Number of responses for C_1 option	Number of responses for C_2 option	Number of responses for C_3 option	Number of responses for C_4 option	Number of responses for C_5 option	Mean	Mode	Median
U_1	23	164	102	44	2	2.52	2	2
U_2	5	12	235	82	1	3.19	3	3
U_3	219	116	0	0	0	1.35	1	1
U_4	25	229	30	21	30	2.41	2	2
U_5	38	79	86	64	68	3.13	3	3
U_6	27	48	155	73	32	3.1	3	3
U_7	108	57	94	52	24	2.48	1	3
U_8	64	64	110	58	39	2.83	3	3
U_9	19	30	142	124	20	3.29	3	3
U_{10}	135	128	43	21	8	1.92	1	2
U_{11}	15	36	84	104	96	3.69	4	4
U_{12}	76	124	79	42	14	2.39	2	2
U_{13}	83	94	1	118	39	2.81	4	2
U_{14}	91	79	99	27	39	2.53	3	2
U_{15}	16	27	118	146	28	3.43	4	4
U_{16}	23	38	173	88	13	3.09	3	3
U_{17}	56	77	150	48	4	2.6	3	3

Table 4.5 Total number of choices selected as per categories of questions

Boxes	Total responses U_1 to U_5	Total responses U_6 to U_{17}	Total responses (T_i) U_1 to U_{17}
Number of responses for C_1 option	310	713	1023
Number of responses for C_2 option	600	802	1402
Number of responses for C_3 option	453	1248	1701
Number of responses for C_4 option	211	901	1112
Number of responses for C_5 option	101	356	457
Total	1675	4020	5695

And $\sum_{i=1}^{5} P(C_i/U_j) = 1$, for all j

Table 4.6 represents the conditional probability of responses C_1, C_2, C_3, C_4 and C_5 given U_j.

Observations

- In this survey, 65.4% male and 34.6% female faculties have given responses.
- This survey is given by highly qualified people, including 70.1% masters and 24.5% doctorate from different disciplines.

Table 4.6 Conditional probability of responses

Boxes	Prob. of C_1 given U_i	Prob. of C_2 given U_i	Prob. of C_3 given U_i	Prob. of C_4 given U_i	Prob. of C_5 given U_i
U_1	0.069	0.49	0.304	0.131	0.006
U_2	0.015	0.036	0.701	0.245	0.003
U_3	0.654	0.346	0	0	0
U_4	0.075	0.684	0.09	0.063	0.09
U_5	0.113	0.236	0.257	0.191	0.203
U_6	0.081	0.143	0.463	0.218	0.096
U_7	0.322	0.17	0.281	0.155	0.072
U_8	0.191	0.191	0.328	0.173	0.116
U_9	0.057	0.09	0.424	0.37	0.06
U_{10}	0.403	0.382	0.128	0.063	0.024
U_{11}	0.045	0.107	0.251	0.31	0.287
U_{12}	0.227	0.37	0.236	0.125	0.042
U_{13}	0.248	0.281	0.003	0.352	0.116
U_{14}	0.272	0.236	0.296	0.081	0.116
U_{15}	0.048	0.081	0.352	0.436	0.084
U_{16}	0.069	0.113	0.516	0.263	0.039
U_{17}	0.167	0.23	0.448	0.143	0.012

- 46.3% and 21.8% faculty of feel that COVID-19 pandemic situation imparted average and high scales, respectively, on their mental health and wellbeing.
- 32.2% of faculty reported that employers had not provided additional support during a pandemic.
- 85.4% of faculty are satisfied with online teaching during a pandemic.
- 78.5% of faculty agreed that there is no impact of a pandemic on their salaries.
- COVID-19 pandemic impacted all sectors but around 60% of faculty observed that the educational sector is highly impacted.
- More than 50% of faculty think that career shift is difficult in the educational sector during pandemic situations.
- More than 50% of faculties are not overcome family crises due to pandemic situations.
- 87.2% of faculty improved their knowledge and skillsets during this pandemic.
- More than 80% of faculty have balance work-life remarkably during a pandemic.
- Around 85% of faculty agreed that students' learning performance and assessment on a digital platform are unsuitable during this pandemic.

Now, the conditional probability of U_j, given C_i, is defined as

$$P(U_1/C_2) = \frac{P(U_1) \times P(C_2/U_1)}{\sum_{j=1}^{17} P(U_j) \times P(C_i/U_j)};$$

And,

$$\sum_{j=1}^{17} P(U_j/C_i) = 1 \text{ for all } i.$$

Table 4.7 depicts the conditional probability of U_j given C_i.

Observations

- If the performance is an average, then the probability that COVID-19 pandemic situation impacted the respondent's mental health and happiness is 0.09 (9%).
- If the performance is deficient (C_1), then 1% of faculty agree that the COVID-19 pandemics impacted significantly less on the educational sector, and 9% of faculty cannot overcome the calamity during a pandemic.
- 11% of faculty feel that additional support provided by their current employer regarding the family's security is meager.

Comparative interpretations of the Male and female respondents for the survey of the Educational sector

Table 4.8 depicts question-wise an average of the responses of male and female respondents.

The interpretations of the above table are given below.

Table 4.7 Conditional probability of U_j given C_i

Boxes	$P(U_j/C_1)$	$P(U_j/C_2)$	$P(U_j/C_3)$	$P(U_j/C_4)$	$P(U_j/C_5)$
U_1	0.02	0.12	0.06	0.04	0
U_2	0	0.01	0.14	0.07	0
U_3	0.21	0.08	0	0	0
U_4	0.02	0.16	0.02	0.02	0.07
U_5	0.04	0.06	0.05	0.06	0.15
U_6	0.03	0.03	0.09	0.07	0.07
U_7	0.11	0.04	0.06	0.05	0.05
U_8	0.06	0.05	0.06	0.05	0.09
U_9	0.02	0.02	0.08	0.11	0.04
U_{10}	0.13	0.09	0.03	0.02	0.02
U_{11}	0.01	0.03	0.05	0.09	0.21
U_{12}	0.07	0.09	0.05	0.04	0.03
U_{13}	0.08	0.07	0	0.11	0.09
U_{14}	0.09	0.06	0.06	0.02	0.09
U_{15}	0.02	0.02	0.07	0.13	0.06
U_{16}	0.02	0.03	0.1	0.08	0.03
U_{17}	0.05	0.05	0.09	0.04	0.01

Table 4.8 Responses of male and female respondents

Gender	Male	Female
U_1	2.62	2.32
U_2	3.18	3.19
U_4	2.44	2.35
U_5	3.3	2.82
U_6	3.1	3.11
U_7	2.45	2.55
U_8	2.95	2.61
U_9	3.19	3.47
U_{10}	1.88	2.01
U_{11}	3.7	3.66
U_{12}	2.31	2.53
U_{13}	2.96	2.52
U_{14}	2.47	2.66
U_{15}	3.36	3.55
U_{16}	3.05	3.16
U_{17}	2.58	2.66

- The average age of males is more than females.
- The average qualification of male and female respondents is approximately equal. Moreover, their job profiles are also identical. However, the average experience of males is more than 15 years, and that of females is less than 15.
- COVID-19 pandemic situation impacted more on mental health and wellbeing of both and feel that additional support provided by their current employer in concern of security of employee and their families is low.
- During a pandemic, the job security of males is hampered more than females. Females are more satisfied with teaching on digital platforms as compared with males.
- The average impact of salaries from the employer is more on a female as compared with males.
- Both respondents agree that COVID-19 pandemic impacted high on the educational sector.
- Females are more satisfied than males with arrangements made by the employer to conduct duties during a pandemic.
- During COVID-19, pandemic males are thought more about career shift as compared with females.
- Both females and males agree that the family crisis increased, and they are trying to come out of it, but females are tackling this situation more comfortably than males.
- There is a reasonable improvement in both respondents' skill sets and knowledge during this pandemic, but females upgraded more than males.
- Females have to manage good work-life balance than males during this COVID-19 pandemic.
- During a pandemic, there is an average learning performance and assessment of students on digital platforms.

Comparative interpretations of the respondents concerning job profiles for the survey of the Educational sector

Table 4.9 depicts question-wise an average of the responses concerning job profiles. The interpretations of the above table are given below.

- COVID-19 pandemic impacted less on Principles/Directors/Deans' mental health and wellbeing than others.
- Supporting staff agree that employers provided good support and security during a pandemic, but others rate it as moderate.
- Every respondent acknowledges that during pandemics, their job security hampers averagely.
- Only Principles/Directors/Deans admitted that their teaching was moderate on this digital platform, and all remaining respondents rate it as well.
- Every respondent concludes that there is a notable impact of a pandemic on salaries during a pandemic.
- Principles/Directors/Deans acknowledge that COVID-19 pandemic impacted very high on educational sector, whereas remaining respondent rate it as high.

Table 4.9 Responses concerning job profiles

Job Profile	Prin/Director	Professor	Asso Professor	Asstt Prof	Supporting
U_1	3.37	3.05	3.1	2.27	2.64
U_2	3.6	3.71	3.57	3.12	2.4
U_3	1.2	1.43	1.3	1.38	1.16
U_5	4.3	4.29	4.07	2.78	2.88
U_6	2.73	2.9	3.23	3.19	2.76
U_7	2.93	2.86	2.27	2.34	3.16
U_8	2.27	2.38	3	2.94	2.72
U_9	2.97	3.33	3.4	3.34	3.04
U_{10}	1.93	2.38	1.87	1.84	2.32
U_{11}	4.1	3.14	3.73	3.69	3.6
U_{12}	2.73	2.57	2.17	2.29	2.92
U_{13}	2.4	2.43	2.83	2.91	2.64
U_{14}	2.77	2.48	2.8	2.5	2.32
U_{15}	3.7	3.62	3.33	3.43	3
U_{16}	3.4	3.43	3.3	3	3.04
U_{17}	2.6	2.62	2.53	2.61	2.64

- An arrangement made by the employer for the conduction of desired duties is rated well by high-profile employees.
- Principles/Directors/Deans agreed less for career shift during the pandemic as compared with others.
- Every respondent agrees that family crises increased during a pandemic, and they are trying to come out of it.
- During this pandemic, there is reasonable improvement in the skill sets and knowledge of all respondents.
- Every respondent has managed an excellent work-life balance during this COVID-19 pandemics. However, Principles/Directors/Deans/Professors have balanced more work-life balance than others.
- During a pandemic, there is an average learning performance and assessment of students on digital platforms.

Comparative analysis of U_6 and U_{14}

Table 4.10 presents the comparative study of the impact of COVID-19 on mental health, wellbeing and family crisis, to what extent an individual can overcome those.

The interpretations of the above table are as follows.

- 0.3% of people have a very high impact of COVID-19 pandemic on their health, wellbeing and family crisis increased in their family but they confidently overcome it.

Table 4.10 Impact of COVID-19 on mental health

$U_6 \backslash U_{14}$	5	4	3	2	1	Total
5	1	2	11	14	4	32
4	7	10	21	24	11	73
3	22	9	52	32	40	155
2	7	6	7	6	22	48
1	2	0	8	3	14	27
Total	39	27	99	79	91	335

- The percentage of people having a very high impact of COVID-19 pandemic on their health and wellbeing, but there is no family crisis is 1.19%.
- 0% of people have a shallow impact of COVID-19 pandemic on their health and wellbeing and family crisis increased but on the verge of recovery.
- 11.94% of people have an average impact of COVID-19 pandemic on their health, wellbeing and no family crisis.

Comparative Analysis of U_1, U_6 and U_{14}

Table 4.11 depicts the comparative study of age group, an effect of COVID-19 pandemic on mental health, wellbeing and family crisis.

The interpretations of the above table are as follows.

- In the age group 20–29 years, there are 0.6% of averagely impacted on their mental health and overcome family crises.
- There are approximately 0% of people having age more than 50 years and this pandemic impacted very high on their mental health and wellbeing but no family crisis.
- Compared with people in the age group 30–49 years, other people (age groups 20–29 and more than 50) feel that there is no family crisis during a pandemic.

Table 4.11 Comparative analysis of U_1, U_6 and U_{14}

$U_6 \backslash U_{14}$ ↓	Age 20–29 years					Age 30–49 years					Age 50 and more than 50 years				
	5	4	3	2	1	5	4	3	2	1	5	4	3	2	1
5	0	0	0	1	1	1	2	9	10	4	0	0	1	3	0
4	0	0	4	3	1	4	10	15	18	8	3	0	2	3	2
3	2	1	2	0	3	18	8	44	30	35	2	0	6	2	2
2	0	0	1	0	1	6	2	5	5	16	1	4	1	1	5
1	0	0	1	1	1	1	0	5	2	8	1	0	2	0	5

4.6 Machine Learning-Based Analysis and Visualization

ML approach is used for the visualization and analysis of survey data. The dataset is in the form excel sheet; however, it is in the textual form, it is converted into the numerical form using encoding. There were no missing values in the dataset as each question in the survey is kept mandatory. Analysis of survey data is done using a *k*-mean algorithm as it is an unlabeled dataset; python language is used for implementing the k-mean algorithm. As discussed in the previous section, few factors are influenced majorly in this pandemic, i.e., mental health, salary deductions, job security, career shift, etc. Analysis of such a few questions is presented in this section.

Feature U_6: Mental Health

Question: How much COVID-19 pandemic situation impacted your mental health and wellbeing?

As per the survey carried out in the education sector, there was no concrete mechanism to take care of mental health during this pandemic. Figure 4.1(i), (ii) and (iii) shows the graphs of mental health impact vs age, experience and job profile, respectively. The clustering of the employees is done based on their job profile and impact mental health on them; Fig. 4.1(iv) shows the clusters that are formed. In the survey, mixed reactions are received, particularly employees in the Principles/Directors/Deans cadre who have experienced a minor impact on their

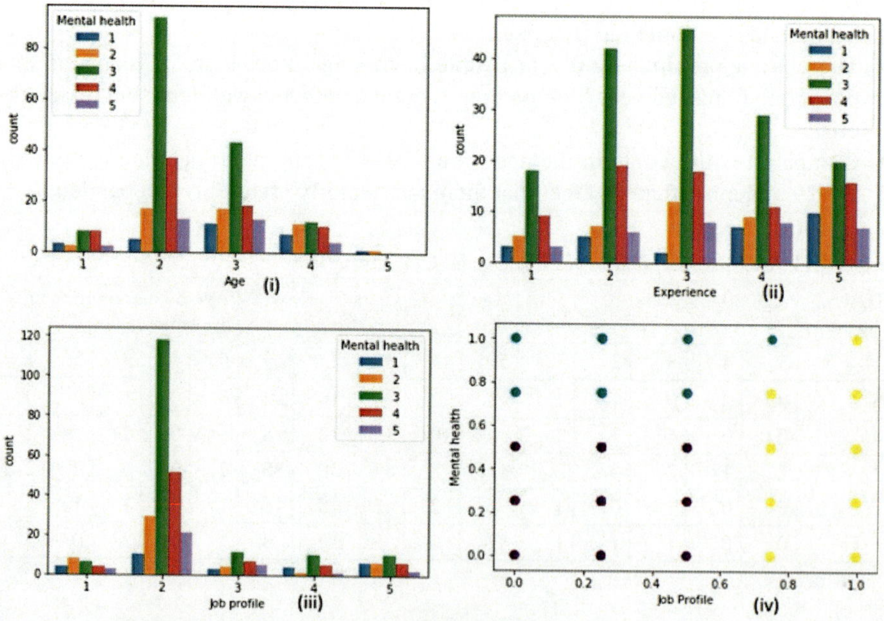

Fig. 4.1 Analysis of COVID-19 on mental health

mental health and wellbeing. It can also be observed that there was more impact on employees' mental health in the beginning phase of their profession.

Given this, it is recommended that there should be some policies in place to ensure the good mental health of employees for all cadres; in turn, it will also improve efficiency and productivity in the long term. The normative inference of this analysis shows that the mental health of maximum employees is likely to be hampered to a large extent with the continuation of this pandemic.

The mathematical modeling and its analysis also present similar observations and are presented in Sect. 4.4.

Feature U_7: Security provided by the employer

Question: Rate additional support provided by your current employer regarding the security of you and your family.

It is employer-centric question. In this pandemic, support by the employer to their employees played a significant role in the lifestyle of employees. Given this, the question is formed to know how much the organizer has supported their employees during the COVID-19 pandemic. The analysis was done based on three important factors like age, experience and employee job profile.

Age

The plot presented in Fig. 4.2(i) shows the behavior of employee's responses from the different age groups toward the security measures taken by employers during this pandemic period. The machine learning analysis of these responses shows a

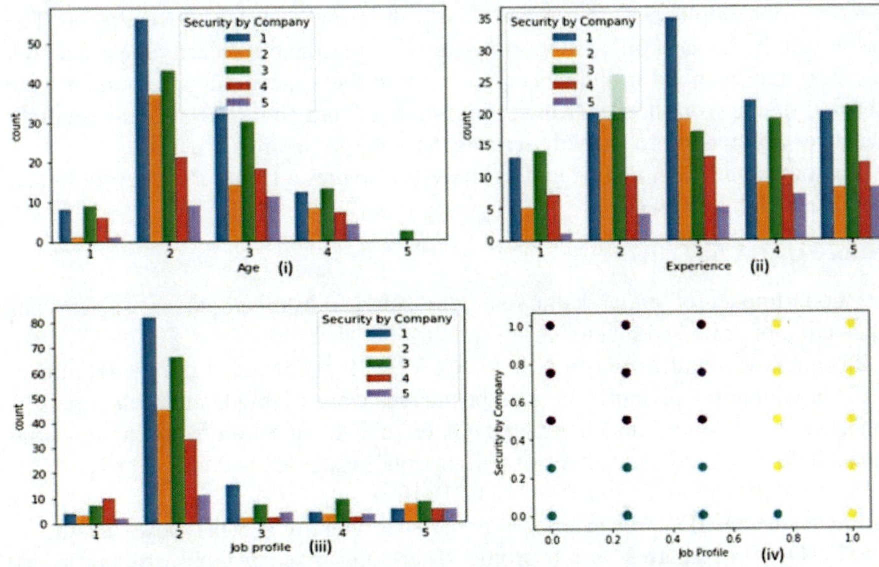

Fig. 4.2 Analysis of security provided by the employer during COVID-19 pandemic

mixed reaction for various measures initiated and deployed by the employer during this pandemic. The analysis shows that employees from major age groups state that the organization took no adequate security measures. However, there is also mixed reaction from employees of all age groups regarding reasonable security measures.

To summarize, there was a lack of security measures taken by the employer for employees and their families. **Given this, it is recommended that adequate security measures will help to improve the productivity of employees.**

Experience

The analysis of employee's experience vs security measures was also carried out in this survey to evaluate the impact of the COVID-19 pandemic. The analysis shows that the realization of security measures was varied for the employees of different experience groups. It is also observed that the policies were different for different experience groups. Figure 4.2(ii) shows that employees with more experience favor no adequate security measures deployed in the organization during this pandemic. **The overall analysis shows that irrespective of employees' experience, marginal security measures were available during this pandemic.**

Job profile

The analysis concerning employee job profiles is also done. The plot in Fig. 4.2(iii) shows that security measures are varied for different profiles. The clustering of the employees is done based on their job profile and security support provided by the organization. Figure 4.2(iv) shows the clusters formed; from the clusters, we can say that the higher job profile states that security measures were adequate while lower position employees are in favor of suitable security measures were taken. The reasons for the lower profile (supporting staff) for reasonable security measures that they were rarely called to the workplace, and all the duties assigned to them were executed online from home. **It is recommended from the analysis that security measures are needed to provide irrespective of job profiles.**

The mathematical modeling and its analysis also present similar observations and are presented in Sect. 4.4.

Feature U_{10}: A pay cut

Question: Impact of a pandemic on your salaries from employer, select your satisfaction of scale given below.

It is employer-centric question. In the COVID-19 pandemic, a significant impact has been seen in the financial domain, businesses were affected, and each organization's economic condition is hampered. Given this, the question regarding financial impact is formed, and an analysis of participants' responses is done.

Figure 4.3(i) shows the impact of COVID-19 on salaries vs age, Fig. 4.3(ii) shows the impact of COVID-19 on salaries vs experience, and Fig. 4.3(iii) shows the impact of COVID-19 on salaries vs job profile. There are mixed responses about salary deduction/pay cut during a pandemic. However, overall analysis from Fig. 4.3(i) is that all age groups' primary responses negatively impact salaries. The analysis

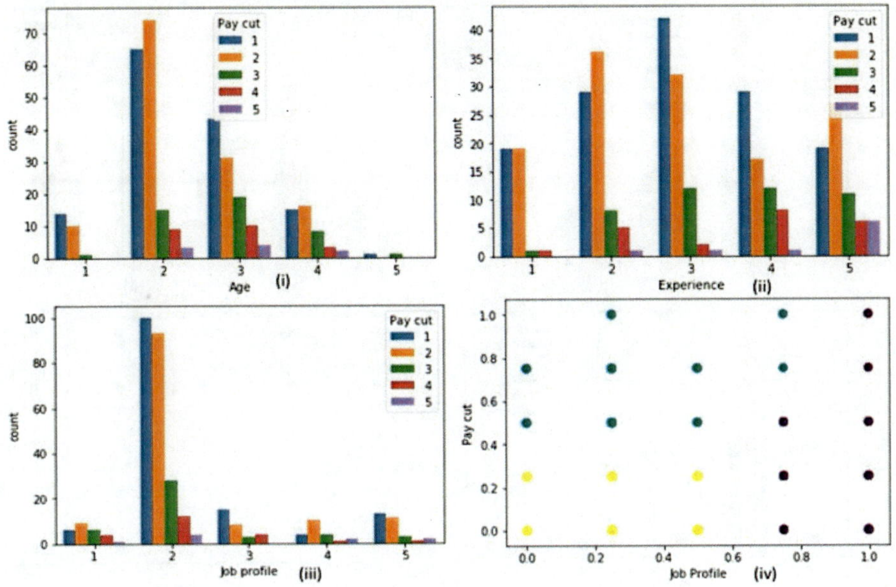

Fig. 4.3 Analysis of COVID-19 impact of salary

was done based on the experience and job profile of employees in this analysis. Also, mixed responses are there; however, most participants stated that there were salary deductions in this pandemic. The clustering of the employees is done based on their job profile and pay cut by the organization; Fig. 4.3(iv) shows the clusters that are formed. **Concisely, we can conclude that there is a negative impact on salaries irrespective of experienceand job profile**. Due to this pandemic, the flow of students to the campus has stopped entirely and affected the flow of money thoroughly, which used to come from a hostel, mess, etc.? In the sequel, the mindset of employers has been completely changed and affected the salary component (irregularities, uncertainties, cut in the pay). Normative inference of this analysis also shows that the same can also be continued until the entire situation gets resumed.

The mathematical modeling and its analysis also present similar observations and are presented in Sect. 4.4.

Feature U_{12}: Arrangements

Question: Are you satisfied with arrangements made by the employer for the conduction of your duties in the pandemic situation?

It is employer-centric question. It was tough to make everything online in the education sector, and there were various hurdles like platforms are new to all stakeholders of these fields, they require good connectivity of the Internet, traditional teaching is very much different from online teaching. Moreover, these online platforms are not free to use; organizations needed to pay more for online teaching. Teachers and

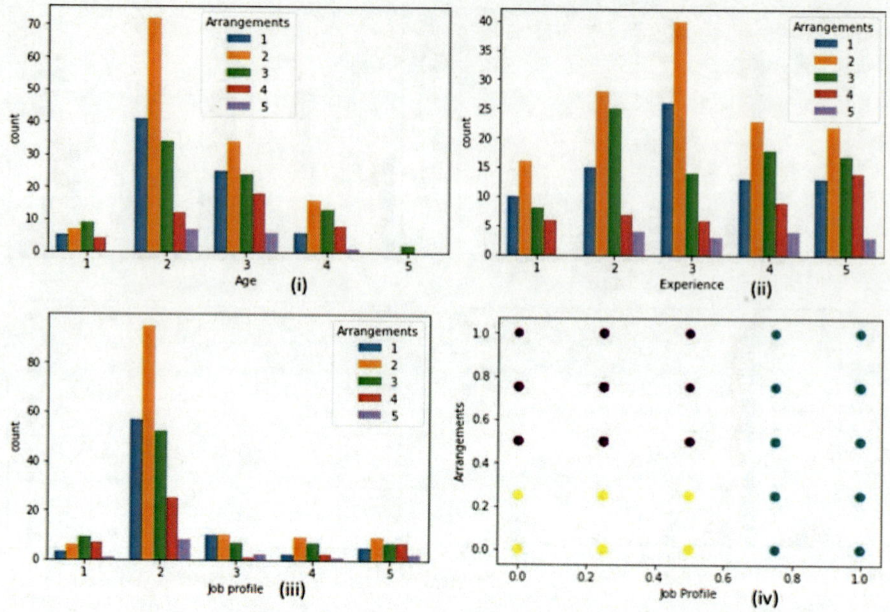

Fig. 4.4 Analysis of arrangements made by the employer during the COVID-19 pandemic

students need the training to use these platforms. Given this, the question is formed, and responses are analyzed.

Age

Figure 4.4(i) shows the arrangements done by employer vs age. From the plots, it is observed that younger age groups are not satisfied with the arrangements made by the employer. The older age groups are stating that adequate arrangements have been made for conducting their duties.

Experience

The analysis is also done based on the experience of employees, as shown in Fig. 4.4(ii); the analysis is similar to age. The less experience (junior) participants are not satisfied with arrangements made by the employer. However, higher experience staff are stating that the employer took adequate arrangements to care.

Job profile

Figure 4.4(iii) shows the analysis of the employee's employer vs job profile. The clustering of the employees is done based on their job profile and pay cut by an organization; Fig. 4.4(iv) shows the clusters that are formed. Analysis shows that the different arrangements are made there for different job profiles. The lower profile participants are not happy with arrangements made by the employer, whereas higher job profile participants stated that the employer follows the suitable arrangements. This inference concludes that adequate arrangement needs to be deployed by employers to enable fear-free work culture even in the pandemic period, which will surely help enhance employees' productivity.

The mathematical modeling and its analysis also present similar observations and are presented in Sect. 4.4.

Feature U13: Career Shift

Question: Is this pandemic situation made you think about a career shift and express your thought below?

In the next part of this analysis, the job profile vs career shift plot is considered for ML-based exploration. Figure 4.5(i), (ii) and (iii) shows the graphs of career shift vs age, experience and job profile, respectively. The clustering of the employees is done based on their job profile and career shift; Fig. 4.5(iv) shows the clusters that are formed. The results show that the more than average employees are in line with their career shift. In addition to this, it is also observed that this trend of career shift can be seen majorly among the employees of junior level. However, employees in the cadre of Principles/Directors/Deans are reluctant toward their career shift. This is an essential outcome of ML-based analysis, which shows that junior employees are more concerned and worried about their existing employment and looking for a career shift.

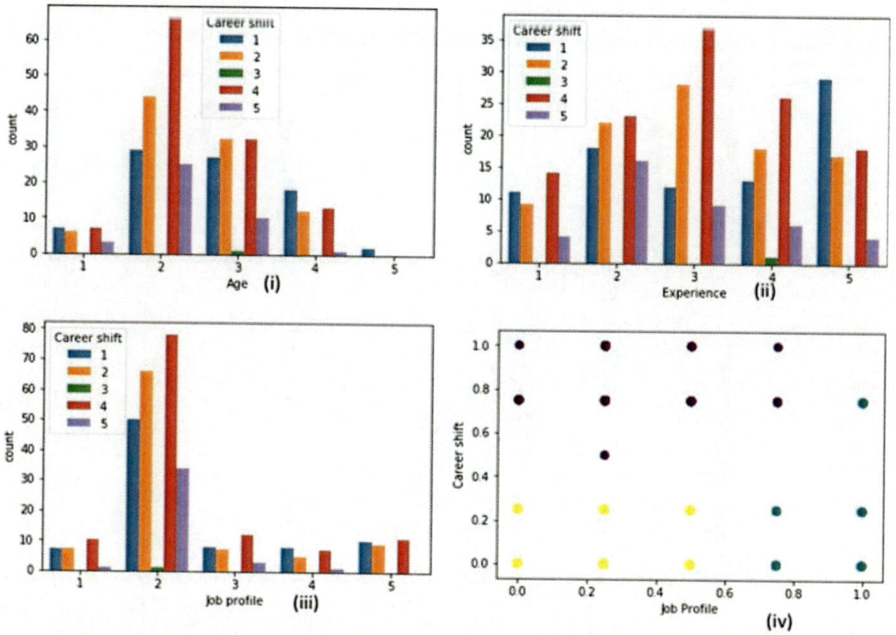

Fig. 4.5 Analysis of COVID-19 pandemic on a career shift

The mathematical modeling and its analysis also present similar observations and are presented in Sect. 4.4.

Feature U16: work-life balance

Question: How much you can make work-life balance in this ongoing situation rate on the below scale.

It was challenging to manage work and family life in this pandemic as duty timings; duty platforms were different. To know the impact of COVID-19 on human lifestyle and how they have balanced work and life. Given this, the question is formed based on how employees make the balance between work and life.

Figure 4.6(i) shows how various age groups are handled work and life during this pandemic. Figure 4.6(ii) shows the responses of employees with varied years of experience. Figure 4.6(iii) shows how employees at different job profiles have handled work-life balance. The clustering of the employees is done based on their job profile and career shift; Fig. 4.6(iv) shows the clusters that are formed. The high-profile employees, including the Principal/Director/Dean, state that there was better work-life balance during this pandemic period. A positive response from high-profile employees is higher pay, position, fewer liabilities, etc.

The question is analyzed based on the age, experience and job profile of the employees. **The analysis shows that the significant response of all age groups, experienceand job profilesis that it was not too easy and difficult to balance**

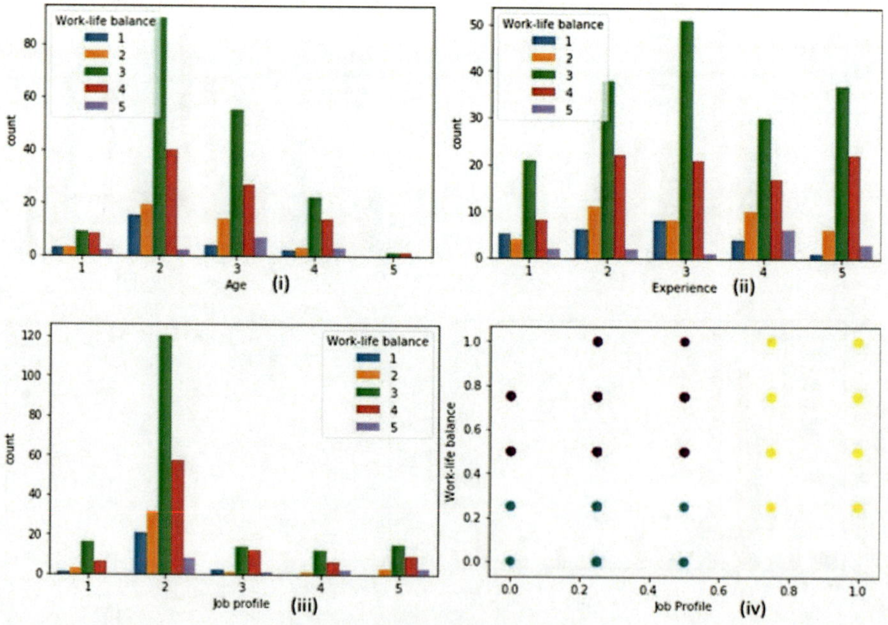

Fig. 4.6 Analysis of COVID-19 pandemic on work-life balance

between work-life during a pandemic. The overall response is that it was moderate to handle life and work during this pandemic.

The mathematical modeling and its analysis also present similar observations and are presented in Sect. 4.4.

4.7 Summary

The uncertain occurrence of the pandemic COVID-19 and its global peak has affected stakeholders of the education sector, i.e., teachers, students, employers, book printing firms, etc. The main factors affected are mental health, loss of salary, job security, career changes, etc. This chapter presented the analysis of the impact of this pandemic on the employees in the education sector. The main contributions presented in this chapter are mathematical modeling and ML examination of the data collected from the survey. The entire dataset will be divided into three categories: demographic, employee-centric and employer-centric information. The visual representation of the analysis will be helpful to the readers for better understanding as well as decision making and recommendations. The conclusions are drawn from mathematical modeling and machine learning-based analysis of education dataset. It is seen that there is a negative impact on salaries irrespective of experience and job profile.

It is also observed that there was more impact on the mental health of employees who are in the beginning phase of their profession. In view of this, it is recommended that there should be some policies in place in order to ensure the good mental health of employees for all cadres, adequate arrangement needs to be deployed by the employer to enable fear-free work culture even in the pandemic period, in turn, it will also improve on efficiency and productivity in long term. The key observation from the analysis for the most important and potential parameters indicates that better policies will be designed and deployed to deal with such situations in the future. The better analysis can be done with the large and balanced dataset, in the survey dataset is in limited size that can be increased with the time and by covering engineering faculties of various states in the India.

References

1. Xiong, J., Lipsitz, O., Nasri, F., Lui, L.M., Gill, H., Phan, L., Chen-Li, D., Iacobucci, M., Ho, R., Majeed, A. and McIntyre, R.S.: Impact of COVID-19 pandemic on mental health in the general population: a systematic review. J. Affective Disorders (2020)
2. Banerjee, D.: The impact of Covid-19 pandemic on elderly mental health. Int. J. Geriatr. Psychiatry (2020)
3. Dey, N., Mishra, R., Fong, S.J., Santosh, K.C., Tan, S., Crespo, R.G.: COVID-19: Psychological and psychosocial impact, fear, and passion. Digital Government: Res. Pract. **2**(1), 1–4 (2020)
4. Sahoo, P., Ashwani: COVID-19 and Indian economy: impact on growth, manufacturing, trade and MSME sector. Glob. Bus. Rev. **21**(5), 1159–1183 (2020)

5. McKibbin, W., Fernando, R.: The economic impact of COVID-19. Econ. Time COVID-19 **45**(10.1162) (2020)
6. He, H., Harris, L.: The impact of Covid-19 pandemic on corporate social responsibility and marketing philosophy. J. Bus. Res. **116**, 176–182 (2020)
7. Tanrıvermiş, H.: Possible impacts of COVID-19 outbreak on real estate sector and possible changes to adopt: a situation analysis and general assessment on Turkish perspective. J. Urban Manage. **9**(3), 263–269 (2020)
8. Siche, R.: What is the impact of COVID-19 disease on agriculture? Sci. Agropecu. **11**(1), 3–6 (2020)
9. Kuhfeld, M., Soland, J., Tarasawa, B., Johnson, A., Ruzek, E., Liu, J.: Projecting the potential impact of COVID-19 school closures on academic achievement. Educ. Res. **49**(8), 549–565 (2020)
10. Ferrel, M.N., Ryan, J.J.: The impact of COVID-19 on medical education. Cureus **12**(3) (2020)
11. Abidah, A., Hidaayatullaah, H.N., Simamora, R.M., Fehabutar, D., Mutakinati, L.: The impact of covid-19 to Indonesian education and its relation to the philosophy of "merdeka belajar." Stud. Philos. Sci. Educ. **1**(1), 38–49 (2020)
12. Pokhrel, S., Chhetri, R.: A literature review on impact of COVID-19 pandemic on teaching and learning. Higher Educ. Future **8**(1), 133–141 (2021)
13. Burgess, S., Sievertsen, H.H.: Schools, skills, and learning: the impact of COVID-19 on education. VoxEu.org **1**(2) (2020)
14. Haleem, A., Javaid, M., Vaishya, R., Deshmukh, S.G.: Areas of academic research with the impact of COVID-19. Am. J. Emerg. Med. **38**(7), 1524–1526 (2020)
15. Sutherland, K., Chessman, J., Zhao, J., Sara, G., Shetty, A., Smith, S., Went, A., Dyson, S., Levesque, J.F.: Impact of COVID-19 on healthcare activity in NSW, Australia. Public Health Res. Pract. **30**(4), e3042030
16. Quezada, R.L., Talbot, C., Quezada-Parker, K.B.: From bricks and mortar to remote teaching: a teacher education program's response to COVID-19. J. Educ. Teach. **46**(4), 472–483 (2020)
17. Assunção Flores, M., Gago, M.: Teacher education in times of COVID-19 pandemic in Portugal: national, institutional and pedagogical responses. J. Educ. Teach. **46**(4), 507–516 (2020)
18. Bhapkar, H.R., Mahalle, P.N., Shinde, G.R., Mahmud, M.: Rough sets in COVID-19 to predict symptomatic cases. In: Santosh, K.C., Joshi, A. (eds.) COVID-19: Prediction, Decision-Making, and Its Impacts. Lecture Notes on Data Engineering and Communications Technologies (60), pp. 57–68. Springer, Singapore (2021). ISBN 9789811596810
19. Bhapkar, H.R., Mahalle, P.N., Dhotre, P.S.: Virus graph and COVID-19 Pandemic: a graph theory approach. In: Hassanien, A.E., Dey, N., Elghamrawy, S. (eds.) Big Data Analytics and Artificial Intelligence Against COVID-19: Innovation Vision and Approach. Studies in Big Data, vol. 78. Springer, Cham (2020). https://doi.org/10.1007/978-3-030-55258-9_2
20. Bhapkar, H.R., Mahalle, P.N., Dey, N., Santosh, K.C.: Revisited COVID-19 mortality and recovery rates: are we missing recovery time period? J. Med. Syst. **44**(12), 202 (2020). https://doi.org/10.1007/s10916-020-01668-6

Chapter 5
Influence of COVID-19 on Quality of Work-Life in Information Technology/Software Industries

Abstract Due to home isolation and social isolation, many information technology (IT) /software companies implemented work from home. IT department employees have a positive attitude toward work from home. Employees can use more energy and time to work because employees do not need to deal with highway traffic, flexibility of work timing/ hours, etc. The downside is that few employees may not have facilities, such as a separate workroom, a good Internet connection; they may be disturbed due to family problems. This chapter is an attempt to understand the influence of COVID-19 on IT/software using data analysis and mathematical modeling.

5.1 Introduction

The COVID-19 pandemic has been faced by every country on the globe, and various adverse effects are witnessed by the various sectors due to the pandemic. The social distancing and lockdown strategies are working better for slowing down the spread rate of COVID-19. These strategies made some restrictions on the daily routine of human life and working of various sectors, i.e., construction industries, education, information technology (IT), health care, agriculture, government organization, etc. In this chapter, the impact of the COVID-19 pandemic on IT industries is discussed. IT industries have also been impacted due to this situation; however, the impact is varied from the other sectors. Due to lockdown and social distancing work from home style of working is implemented by many IT companies. It is observed that the strong and wide Internet connection is the prime requirement for the smooth working of IT companies.

The employees of the IT sector are toward the positive side of the work from home. In metropolitan cities traveling is the biggest challenge, especially in the IT hub areas heavy road traffic is observed and due to work from home employees do not need to face heavy road traffic, it results in saving precious time. The employees can work with more energy and time. There is downside also few employees may not have facilities like the separate room to work, strong Internet connection, they may get disturbance due to family issues. In line with this, the chapter focuses on the impact of COVID-19 on the quality of working life of IT employees. The survey

G. R. Shinde et al., *Quality of Work-Life During Pandemic*, Studies in Big Data 100,
https://doi.org/10.1007/978-981-16-7523-2_5

has been conducted to know the opinion of the IT employees regarding the impact of COVID-19 on their work-life. The questionnaire for this survey was designed in consultation with all stakeholders of the IT companies. The questions focus on the support received by the employees from the employers, difficulties faced in work from home. The questionnaire was prepared and survey responses were collected via Google forms. The questionnaire was emailed to employees of IT companies. The employees with different experiences, qualifications and job profiles have participated in the survey. These questions and their responses are then analyzed using mathematics and machine learning tools to draw conclusions.

The main contributions of this study are:

- The preparation of a questionnaire for employees in IT sectors for recording their ongoing experiences during this COVID-19 pandemic.
- Preparation of the dataset resulted from the responses collected from the survey initiated for employees in the IT sectors.
- Design a mathematical model to draw statistical inferences, correlation in the form of association rule mining and key observations.
- Machine learning-based analysis and visualization to draw useful inferences and normative interpretations from the dataset.

5.2 Literature Work

The traditional work strategy of IT employees has been disturbed due to the COVID-19 pandemic, and the work from home has its pros and cons. Many researchers have discussed the impact of this pandemic on IT industries. A. Bolhari et al. [1] have conducted survey, 292 IT staff are participated in the survey. Analysis of survey suggests the level of quality of work and mechanism to enhance quality of work. This is done by measuring the quality of the work-life of IT staff and checking the correlation between the quality of work-life and societal characteristics among them. Igbaria et al. [2] presented the analysis of survey that conducted to understand parameters that are important for quality of work-life of employees. In the survey, 464 professionals and managers have participated. The analysis presents factors that affecting the quality of work, one of the parameter is job involvement. There is a low, moderate and high level of job involvement, a high level of involvement may give beneficial results to enhance the quality of work. Rohini et al. [3] investigated the impact of work-life balance and different facts. In few IT industries, it becomes a big challenge to manage professional responsibilities as well as commitments and personal life activities. Leading employers take their work-life positively and that is key ingredient of a successful business strategy. Vijay et al. [4] have surveyed IT employees in three metropolitan cities Bangalore, Chennai and Coimbatore. The data was captured with respect to specific factors like control at work, general well-being, homework interface, job career satisfaction, stress at work and working conditions. Statistical analysis is done using Chi-square statistics to study the association between employee's overall opinion on the quality of work and their work productivity. Bharathi et al. [5] presented the analysis of quality of work-life for woman employees. The survey of woman employees in the Bangalore City working in Information Technology-Information Technology Enabled Services (IT-ITES). This firm is helping women to establish their names in the competitive globe by being good homemaker. With increased responsibilities, stress automatically gets increased, they need to apply different coping methods to come out of the stress. The study revealed that the distractive positive method of coping style was most widely used by women employees. In a pandemic situation, software engineers were forced or needed to work remotely from home. This affected the work efficiency, their wellbeing and productivity. Russo et al. [6] analyzed that it is very difficult to understand how they are maintaining their wellbeing, psychological and social variables. Russo [7] investigated the impact of work from home and its restrictions due to the COVID-19 pandemic situation on an individual level among software engineers. In COVID-19 pandemic, governments have forced restrictions for citizens to lower down the virus spread rate. This affected the psychological, social, situational and physiological factors that have been associated with the wellbeing or productivity. Anxiety, distractions, coping strategies, psychological and physical needs, office setup, stress and work motivation are the factors that affect work efficiency or productivity. Ford et al. [8] survey led to powerful narratives and that provided insights about the way remote work can evolved in the future of pandemic. Software engineers are being migrated from their traditional working style to work from home because of this COVID-19 pandemic situation. This pandemic has affected the software engineers and their productivity. There is the benefit of being close to the family members, and some have faced the problem due to work from home as they needed to share the working place with family members. Bhapkar et al. [9] given mathematical model to reduce symptoms of COVID-19 by using rough set theory approach. Bhapkar et al. [10] presented any types of viruses by using graphs and defined growth rates of viruses in terms of the corona product of graphs.

Bhapkar et al. [11] defined the first generic formulae to measure the severity of any epidemic or pandemic during its period also, known as the progressive recovery rate and progressive mortality rate.

The chapter is organized as follows, the description of data and dataset are discussed in Sect. 5.3, and statistical description of the dataset is presented in Sect. 5.4. The mathematical modeling for the educational dataset is designed and presented in 5.5. The data is analyzed and visualized using ML techniques, presented in Sect. 5.6.

5.3 Dataset and Data Description

A survey was conducted in June 2021 and 131 employees from the education sector with different profiles participated. The questionnaire consists of 22 questions; five of them are for demographic information, and other questions are related to the impact of COVID-19 on the quality of working life. It is intended that survey participants responses to the questions by selecting one of the answers from five response options.

The responses are in the form of fuzzy linguistic variables/values that scale from low–high to make it easier for respondents to complete the survey while having fun with the survey process.

The questions are related to the way employer has supported their employees in this pandemic situation, the up and downside of work from home style of working, the mental stability of employees, etc.; these are presented in Table 5.1.

Table 5.1 Questionnaire

S. No.	Questions
Q_1	What is your age?
Q_2	What is your qualification?
Q_3	What is your gender?
Q_4	What is your job profile?
Q_5	What is your experience?
Q_6	How much COVID-19 pandemic situation impacted on your mental health and wellbeing in this IT world?
Q_7	Extent to which COVID-19 pandemic hampered your job security?
Q_8	Rate additional support provided by your current employer in concern of security of you and your family
Q_9	To what extent your work was satisfactory on this digital platform?
Q_{10}	Impact of pandemic on your salaries from employer, select your satisfaction of scales given below
Q_{11}	COVID-19 pandemic impacted all sectors, how you rate IT industry in this concern
Q_{12}	Are you satisfied with arrangements made by employer for conduction of your duties in pandemic situation?
Q_{13}	Is this pandemic situation made you to think about career shift, express your thought on below scales?
Q_{14}	If the answer of above question is 'Never,' then have you tried to enhance your skills?
Q_{15}	Do you feel online training courses helping to upskill you?
Q_{16}	Do you feel the guided training is best way to learn and enhance skills?
Q_{17}	Do you think because 'Working from Home,' your productivity has increased or decreased?
Q_{18}	Do you feel the socializing with team is missing in current 'Working from Home' situation and it is impacting your social behavior?
Q_{19}	Due to uncertain situation from last year, is family crisis increased and to what extent you are able to overcome on those
Q_{20}	Rate the improvement in your skill sets and knowledge during this pandemic?
Q_{21}	How much you are able to make work-life balance in this ongoing situation rate on below scale
Q_{22}	As per your opinion, what is the assessment of IT professionals on this digital platform?

5.4 Descriptive Statistics

Dataset can be described briefly with statistical information. Statistical information can describe the central tendency of the dataset; i.e., how are the data distributed? How do they scatter from their mean? What is the response given by most participants for a particular characteristic (mode)? What is the mean value of a particular characteristic? The qualitative statistical description of this dataset is presented in Table 5.2. The Pandas library in Python is used to calculate statistical values of each, the features are renamed as U1 to U17, and in further sections of the chapter, these naming conventions are used.

Table 5.2 Statistical values

Demographic/ variables	Std deviation	Mean	Mode	Median	Max	Min
Q_1	0.714172	0.832061	**0**	1	3	0
Q_2	0.536074	1.396947	1	1	3	1
Q_3	0.426652	0.236641	1	0	1	0
Q_4	1.92898	2.954198	0	3	7	0
Q_5	1.237931	1.335878	3	1	4	0
Q_6	1.16974	1.969466	0	2	4	0
Q_7	1.145391	1.282443	2	1	4	0
Q_8	1.230175	2.51145	0	3	4	0
Q_9	1.014284	2.587786	3	3	4	0
Q_{10}	1.103137	1.564885	3	1	4	0
Q_{11}	1.095927	1.641221	1	2	4	0
Q_{12}	1.150812	2.312977	…	2	4	0
Q_{13}	1.12793	1.29771	1	1	4	0
Q_{14}	1.272271	2.450382	4	2	4	0
Q_{15}	0.761755	0.793893	1	1	3	0
Q_{16}	0.79856	0.908397	1	1	3	0
Q_{17}	1.169188	1.900763	3	2	3	0
Q_{18}	0.705319	0.625954	0	0	2	0
Q_{19}	1.331624	1.458015	0	1	4	0
Q_{20}	0.967165	2.183206	2	2	4	0
Q_{21}	1.070394	1.847328	2	2	4	0
Q_{22}	0.763295	2.412214	2	2	4	0

5.5 Mathematical Modeling

Table 5.3 depicts the correlation coefficients between two different datasets.

From the table, the following observations are found.

- There is a high positive correlation between Q_1 and Q_5.
- Q_8 and Q_{12} have moderate positive correlation between them.
- The questions having low positive correlation are as given below Q_1 and Q_2, Q_6 and Q_7, Q_8 and Q_9, Q_8 and Q_{10}, Q_9 and Q_{12}, Q_{10} and Q_{12}, Q_{13} and Q_{15}, Q_{14} and Q_{20}, Q_{17} and Q_{18}, Q_{18} and Q_{21}, Q_{20} and Q_{22}, Q_{21} and Q_{22}.
- Questions Q_6 and Q_8, Q_6 and Q_9, Q_6 and Q_{12}, Q_6 and Q_{21}, Q_7 and Q_8, Q_7 and Q_{12} have low negative correlation.
- There are 211 negligible correlations.

By considering these relations, the Correlation Graph is defined as follows.

Correlation Graph

A graph C is said to be Correlation Graph with vertex set and edge set as follows.

Vertex Set = {Set of questions or observations or points} and.

Edge Set = $\{\{Q_i, Q_j\} / i \neq j \text{ and } r(Q_i, Q_j) \geq 0.3\}$. That is, Q_i and Q_j are adjacent to each other if and only if the correlation coefficients between Q_i and Q_j are greater than or equal to 0.3.

The subgraphs with respect to correlations are given below. Here, we have used $Q_i = I$ to denote vertices in graphs.

1. Subgraph for high positive correlation is given in Fig. 5.1
2. Subgraph for moderate positive correlation is specified in Fig. 5.2
3. Figure 5.3 depicts the subgraph for low positive correlation.
4. Subgraph for low negative correlation is shown in Fig. 5.4:

   By considering all above graphs, the Correlation Graph of the data is given Fig. 5.5.

Weighted Correlation Graph

A Correlation Graph in which weight of each edge is the correlation coefficient between end vertices of that edge is called Weighted Correlation Graph.

In this chapter, our focus mainly is on the following question.

Q6: How much COVID-19 pandemic situation impacted on your mental health and wellbeing in this IT world?

In above graph, 6 is adjacent to vertices 7, 8, 9, 12 and 21. All remaining vertices are connected or disconnected with vertex 6 but their weights are less than 0.3 with vertex 6.

Let us consider the Q_6, Q_7, Q_8, Q_9, Q_{12} and Q_{21} for further analysis of the data. If the focus of study is Q_{12} then consider questions Q_6, Q_7, Q_8, Q_9, Q_{12} and Q_{10}. Let us start with question 6.

Table 5.3 Correlation Coefficients of Datasets

Que. No	Q_2	Q_3	Q_4	Q_5	Q_6	Q_7	Q_8	Q_9	Q_{10}	Q_{11}	Q_{12}	Q_{13}
Q_1	0.376	0.005	0.151	0.786	0.16	0.134	−0.129	0.042	−0.074	−0.009	−0.104	−0.004
Q_2		0.124	0.092	0.215	0.056	0.117	−0.135	0.006	0.047	−0.018	0.009	0.045
Q_3			0.097	−0.093	0.2	0.146	−0.042	−0.022	0.024	0.101	−0.105	−0.1
Q_4				0.29	−0.089	0.023	0.179	0.136	0.175	0.047	0.159	−0.071
Q_5					0.124	0.063	−0.048	0.087	−0.056	0.027	0.044	−0.034
Q_6						0.368	−0.304	−0.302	−0.183	0.171	−0.33	0.228
Q_7							−0.316	−0.184	−0.152	0.167	−0.383	0.256
Q_8								0.442	0.313	−0.12	0.554	−0.26
Q_9									0.244	−0.093	0.428	−0.194
Q_{10}										0.048	0.381	−0.136
Q_{11}											−0.166	0.087
Q_{12}												−0.197
Q_{13}												

Que. No	Q_{14}	Q_{15}	Q_{16}	Q_{17}	Q_{18}	Q_{19}	Q_{20}	Q_{21}	Q_{22}
Q_1	−0.162	−0.008	−0.027	−0.029	−0.034	−0.096	−0.178	−0.134	−0.084
Q_2	0.142	0.108	−0.13	0.027	0.03	−0.073	−0.112	0.053	−0.065
Q_3	0.029	−0.133	−0.049	−0.061	0.041	0.146	−0.031	−0.005	0.1
Q_4	−0.029	0.083	−0.028	0.073	0.016	−0.142	−0.127	0.161	−0.013

(continued)

Table 5.3 (continued)

Que. No	Q_{14}	Q_{15}	Q_{16}	Q_{17}	Q_{18}	Q_{19}	Q_{20}	Q_{21}	Q_{22}
Q_5	−0.024	−0.089	−0.031	0.05	−0.04	−0.103	−0.193	−0.06	−0.099
Q_6	−0.125	0.166	0.005	−0.137	−0.154	0.108	−0.274	−0.372	−0.253
Q_7	−0.109	0.173	−0.03	−0.031	−0.087	−0.01	−0.13	−0.103	−0.196
Q_8	0.004	−0.281	−0.046	0.116	0.134	−0.088	0.153	0.153	0.265
Q_9	0.139	−0.19	0.038	0.218	0.138	−0.127	0.281	0.282	0.291
Q_{10}	−0.002	−0.162	−0.072	−0.04	0.125	0.037	−0.018	0.276	0.123
Q_{11}	0.073	0.141	0.05	−0.01	−0.155	0.077	0.055	−0.06	−0.061
Q_{12}	0.181	−0.154	0.006	0.241	0.107	−0.059	0.121	0.27	0.281
Q_{13}	0.072	0.314	0.013	0.034	−0.062	0.057	0.013	−0.039	−0.162
Q_{14}		0.057	0.109	0.139	0.009	0.241	0.376	0.203	0.195
Q_{15}			0.108	0.003	−0.102	0.018	0.01	0.065	−0.21
Q_{16}				0.015	0.007	0.025	0.022	0.002	−0.076
Q_{17}					0.346	0.069	0.261	0.252	0.175
Q_{18}						−0.005	0.225	0.372	0.103
Q_{19}							0.102	−0.112	0.032
Q_{20}								0.235	0.335
Q_{21}									0.304
Q_{22}									

Fig. 5.1 Subgraph for high positive correlation

Fig. 5.2 Subgraph for high positive correlation

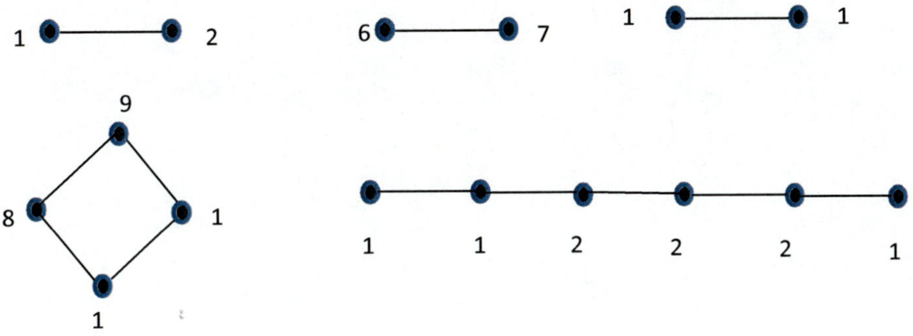

Fig. 5.3 Subgraph for low negative correlation

Fig. 5.4 Subgraph for low negative correlation

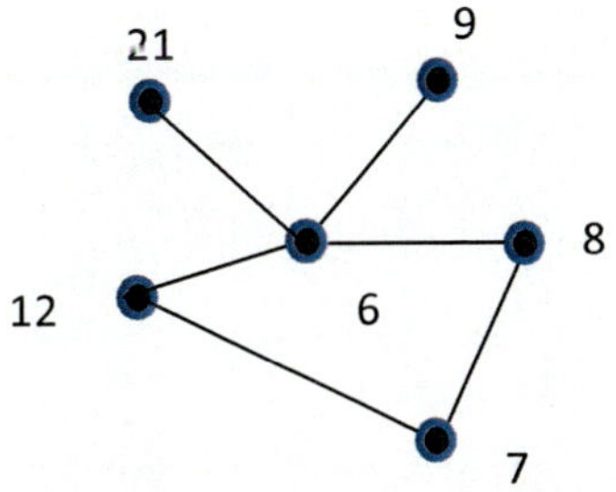

We would like to verify an effect of COVID-19 on mental health wellbeing of an individual in IT sector by considering the above dataset. We have selected six questions and 15 responses from the dataset as a sample case. The mental health of people is described by using the five questions Q_7, Q_8, Q_9, Q_{12} and Q_{21} together with the corresponding scales ordered from 0 to 4, given below.

Therefore, the set of condition attributes is $C = \{Q_7, Q_8, Q_9, Q_{12}, Q_{21}\}$. Fifteen different responses have been sorted by us after due mathematical consideration. The set of decision attributes is $D = \{\text{Yes}, \text{No}\}$, where Yes: COVID-19 is affected on the mental health and wellbeing during pandemic period. No: COVID-19 is not affected on the mental health and wellbeing during pandemic period. The desired information is presented in Table 5.4

From the table, we have,

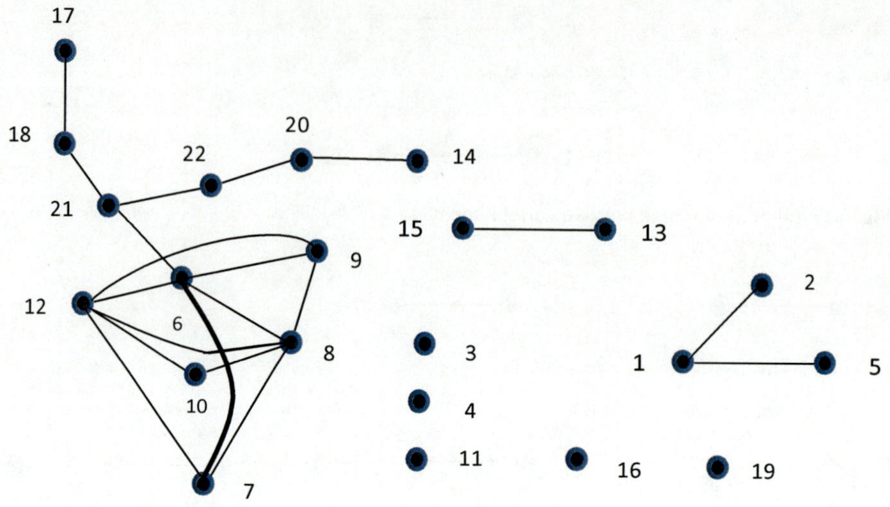

Fig. 5.5 Correlation graph for dataset

Table 5.4 Decision table for IT sector

Condition/responses	Q_7	Q_8	Q_9	Q_{12}	Q_{21}	Decision
R_1	0	2	2	3	2	No
R_2	0	4	3	2	3	No
R_3	1	1	1	0	0	No
R_4	0	3	4	3	2	No
R_5	0	4	1	3	1	Yes
R_6	2	0	2	1	2	Yes
R_7	1	4	1	2	1	Yes
R_8	2	3	3	1	2	Yes
R_9	1	4	4	2	2	Yes
R_{10}	1	4	4	4	2	Yes
R_{11}	3	1	2	0	1	Yes
R_{12}	1	2	3	2	2	Yes
R_{13}	0	2	2	3	2	Yes
R_{14}	1	1	1	4	4	Yes
R_{15}	2	3	4	3	3	Yes

$X_Y = \{R_5, R_6, R_7, R_8, R_9, R_{10}, R_{11}, R_{12}, R_{13}, R_{14}, R_{15}\}$ and $X_N = \{R_1, R_2, R_3, R_4\}$.

The lower approximation and upper approximation of rough set of X_Y and X_N are given below.

$\underline{C}X_Y = \{R_5, R_6, R_7, R_8, R_9, R_{10}, R_{11}, R_{12}, R_{14}, R_{15}\} = X_Y - \{R_{13}\}$.

$\overline{CX_Y} = \{R_1, R_5, R_6, R_7, R_8, R_9, R_{10}, R_{11}, R_{12}, R_{13}, R_{14}, R_{15}\} = X_Y \cup \{R_1\}$ and
$\underline{CX_N} = \{R_2, R_3, R_4\} = X_N - \{R_1\}$.
$\overline{CX_N} = \{R_1, R_2, R_3, R_4, R_{13}\} = X_N \cup \{R_{13}\}$.
Moreover, $Bnd_C(X_Y) = \{R_1, R_{13}\}$ and $Bnd_C(X_N) = \{R_1, R_{13}\}$.

The next step of the rough set analysis of the decision table is construction of minimal subset of independent criteria ensuring the same quality of sorting as whole set C, that means reduces of C. The following are the required reduces in our case.
$RED^1{}_X = \{Q_7, Q_8, Q_9, Q_{21}\}$, $RED^2{}_X = \{Q_7, Q_{12}, Q_{21}\}$, $RED^3{}_X = \{Q_7, Q_8, Q_9\}$.
It is observed that the criteria U_7 is not present in the positive decision.

Case 1: Select $RED^1{}_X$

Suppose we have chosen to reduce $RED^1{}_X$ composed of Q_7, Q_8, Q_9, Q_{21}, that is, job security, additional support by employer, work satisfaction and work-life balance. Now the decision table can be reduced to criteria given in $RED^1{}_X$ is given in Table 5.5.

The decision rules or laws are generated from the reduced decision table and have the following forms as shown in Table 5.6:

Now, consider Table 5.7 for certainty factor and coverage factor of above rules. By the certainty factors of the decision rules, we can conclude the following.

- If the score of Q_7 is 2 or 3, then COVID-19 is affected on the mental health certainly during pandemic period.
- If the score of Q_7 is 2 and the score of Q_8 is 3, then COVID-19 is affected on the mental health certainly during pandemic period. That is, if the job security is low

Table 5.5 Reduced decision table for $RED^1{}_X$

Condition/responses	Q_7	Q_8	Q_9	Q_{21}	Decision
R_1	0	2	2	2	No
R_2	0	4	3	3	No
R_3	1	1	1	0	No
R_4	0	3	4	2	No
R_5	0	4	1	1	Yes
R_6	2	0	2	2	Yes
R_7	1	4	1	1	Yes
R_8	2	3	3	2	Yes
R_9	1	4	4	2	Yes
R_{10}	1	4	4	2	Yes
R_{11}	3	1	2	1	Yes
R_{12}	1	2	3	2	Yes
R_{13}	0	2	2	2	Yes
R_{14}	1	1	1	4	Yes
R_{15}	2	3	4	3	Yes

Table 5.6 Decision rules for $RED^1{}_X$

Rule No. L_i	Rules	Decision (d, Y/N)
1	(Q_7, 2) OR (Q_7, 3)	Y
2	(Q_7, 2) ^ (Q_8, 3)	Y
3	(Q_9, 1) ^ (Q_{21}, 1)	Y
4	(Q_8, 4) ^ (Q_9, 1) ^ (Q_{21}, 1)	Y
5	(Q_7, 2) ^ (Q_{21}, 2)	Y
6	(Q_7, 1) ^ (Q_8, 1) ^ (Q_9, 1)	Y
7	(Q_7, 1) ^ (Q_8, 1) ^ (Q_9, 1)	N

Table 5.7 Certainty and coverage factors for decision rules of $RED^1{}_X$

Rule	Certainty factor	Coverage factor
1	1	(4/11) = 0.36
2	1	0.18
3	1	0.18
4	1	0.18
5	1	0.18
6	0.5	0.09
7	0.5	0.09

and additional support by employer during pandemic is moderate then COVID-19 is certainly affected on the mental health and wellbeing.

- If Q_9 is 1 and Q_{21} is also 1, then COVID-19 is affected on the mental health certainly during pandemic period. That is, if the work satisfaction is low and work-life balance is low during pandemic, then COVID-19 is certainly affected on the mental health and wellbeing.
- If the score of Q_8, Q_9 and Q_{21} is 4, 1 and 1, respectively, then COVID-19 is affected on the mental health certainly during pandemic period.
- If the score of Q_7 and Q_{21} is 2, then COVID-19 is affected on the mental health certainly during pandemic period.
- If the score of Q_7, Q_8 and Q_9 is 1, then the possibility that COVID-19 is affected on the mental health and wellbeing during pandemic period is 50%.

Similarly, we can apply same procedure for remaining all reduces.

5.6 Machine Learning-Based Analysis and Visualization

ML approach is used for the visualization and analysis of survey data. The dataset is in the form excel sheet; however, it is in the textual form; hence, it is converted into the numerical form using encoding. There were no missing values in the dataset as each question in the survey is kept mandatory. Analysis of survey data is done using a k-mean algorithm as it is an unlabeled dataset; python language is used for implementing the k-mean algorithm.

As discussed in the previous section, few factors are influenced majorly in this pandemic, i.e., mental health, salary deductions, job security, productivity, etc. Analysis of such a few questions is presented in this section.

Q6. How much COVID-19 pandemic situation impacted your mental health and wellbeing in this IT world?

As per the survey carried out in the IT sector, there was a lack of mechanisms to take care of mental health during this pandemic. Figure 5.6(i) and (iii) shows the graphs of mental health impact vs job profile, and experience, respectively. The clustering of the employees is done based on their job profile, experience and impact mental health on them; Fig. 5.6(ii) and (iv) shows the clusters that are formed. In the survey, mixed reactions are received, particularly employees in the higher cadre who have experienced a minor impact on their mental health and wellbeing. It can also be observed that there was more impact on employees' mental health in the beginning phase of their profession.

Given this, it is recommended that there should be some policies in place to ensure the good mental health of employees for all cadres; in turn, it will also improve efficiency and productivity in the long term. The normative inference of this analysis shows that the mental health of maximum employees is likely to be hampered to a large extent with the continuation of this pandemic.

Q7. Extent to which COVID-19 pandemic hampered your job security?

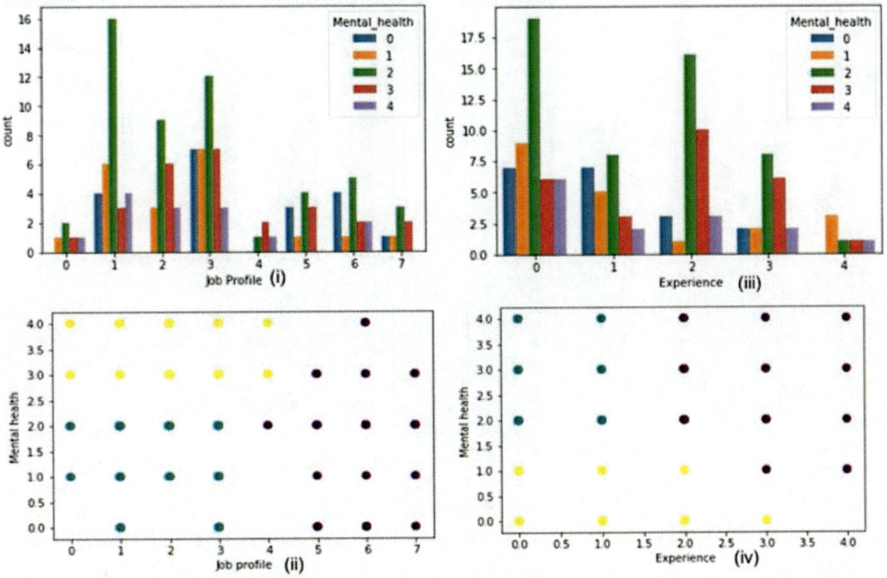

Fig. 5.6 Analysis of impact of Covid-19 on mental health

The job security parameter is considered for ML-based exploration. Figure 5.7(i) and (iii) shows the graphs of job security vs job profile and experience, respectively. The clustering of the employees is done based on their job profile, experience and job security; Figure 5.7 (ii) and (iv) shows the clusters that are formed. The results show that the employees do not feel insecure about the job in this pandemic. In addition to this, it is also observed that secure feelings about the job can be seen majorly among the employees of junior level. This is an essential outcome of ML-based analysis, which shows that employees of the IT sector are not worried about their existing employment.

Q10. Impact of pandemic on your salaries from employer, select your satisfaction of scales given below

In the COVID-19 pandemic, a significant impact has been seen in the financial domain, businesses were affected, and each organization's economic condition is hampered. Given this, the question regarding financial impact is formed, and an analysis of participants' responses is done.

Figure 5.8(i) shows the impact of COVID-19 on salaries vs job profile, and Figure 5.8(iii) shows the impact of COVID-19 on salaries vs experience. The clustering of the employees is done based on their job profile, experience and salaries given by the organization; Figure 5.8(ii) and (iv) shows the clusters that are formed. There are mixed responses about salary deduction/pay cut during a pandemic. Also, mixed responses are there; however, most participants stated that there were salary

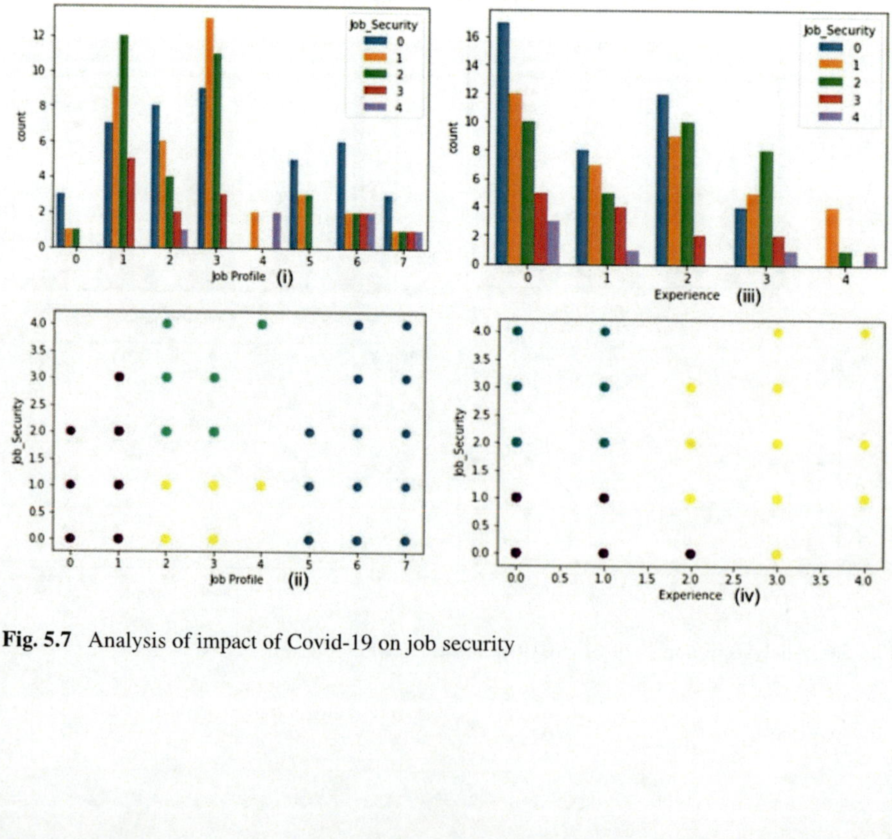

Fig. 5.7 Analysis of impact of Covid-19 on job security

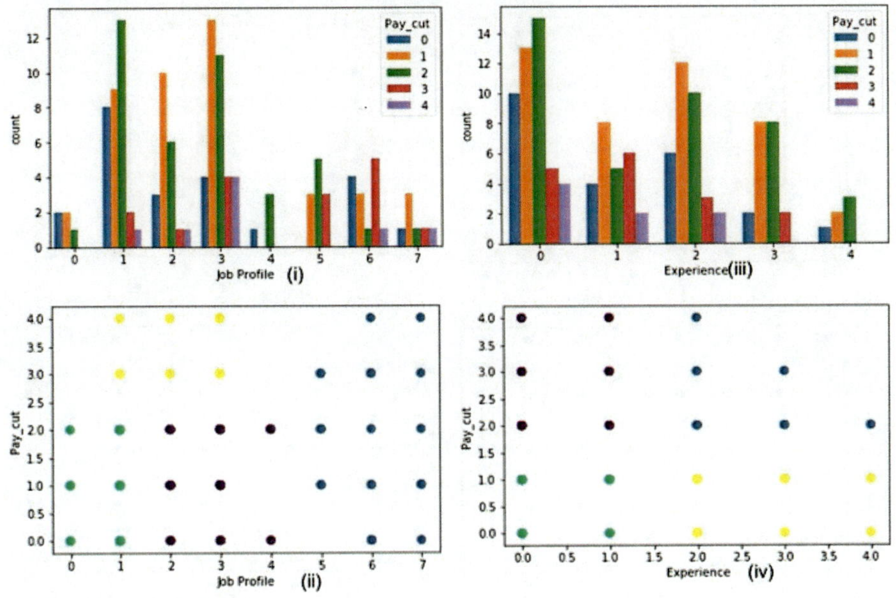

Fig. 5.8 Analysis of impact of Covid-19 impact of salary

deductions in this pandemic. **Concisely, we can conclude that there is a negative impact on salaries irrespective of experienceand job profile**.

Q12. Are you satisfied with arrangements made by employer for conduction of your duties in pandemic situation?

Figure 5.9(i) shows the arrangements made by employer vs job profile, and Figure 5.9(iii) shows the arrangements made by employer vs experience. The clustering of the employees is done based on their job profile, experience and arrangements done by the employer; Figure 5.9(ii) and (iv) shows the clusters that are formed.

Analysis shows that the different arrangements are made there for different job profiles. The lower profile participants are not happy with arrangements made by the employer, whereas higher job profile participants stated that the employer follows the suitable arrangements. This inference concludes that adequate arrangement needs to be deployed by employers to enable fear-free work culture even in the pandemic period, which will surely help enhance employees' productivity.

Q.21 How much you are able to make work-life balance in this ongoing situation rate on the below scale

It was challenging to manage work and family life in this pandemic as duty timings. To know the impact of COVID-19 on human lifestyle and the way they have balanced work and life. Given this, the question is formed based to know the way employees make the balance between work and life.

Figure 5.10(i) shows employees with various job profiles have handled work and life during this pandemic. Figure 5.10(iii) shows the responses of employees

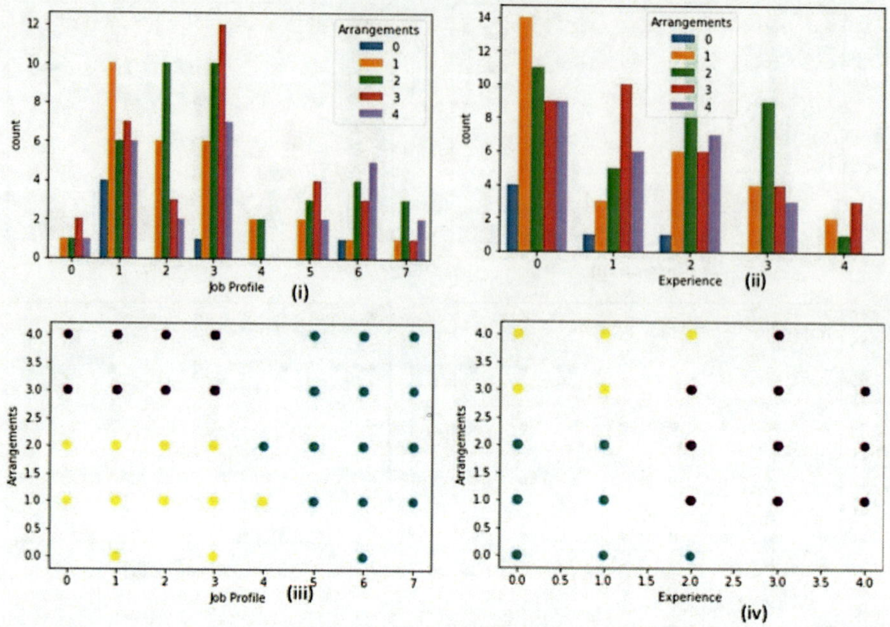

Fig. 5.9 Analysis of arrangements made by the employer during the Covid-19 pandemic

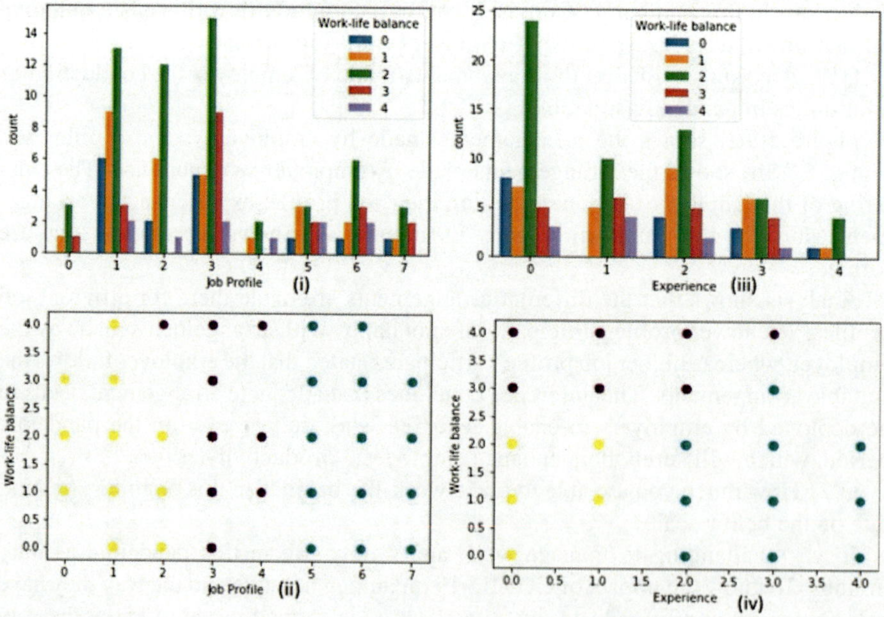

Fig. 5.10 Analysis of impact of Covid-19 pandemic on work-life balance

with varied years of experience. The clustering of the employees is done based on their job profile and experience; Figure 5.10(ii) and (iv) shows the clusters that are formed. The high-profile employees, including the state that there was better work-life balance during this pandemic period. **The analysis shows that the significant response of all experience and job profiles is that it was not too easy and difficult to balance between work-life during a pandemic. The overall response is that it was moderate to handle life and work during this pandemic**.

5.7 Summary

In this chapter, the impact of the COVID-19 pandemic on IT industries is discussed. IT industries have also been impacted due to this situation; however, the impact is varied from the other sectors. It is observed that the strong and wide Internet connection is the prime requirement for the smooth working of IT companies. The employees of the IT sector are toward the positive side of the work from home. In metropolitan cities, traveling is the biggest challenge, especially in the IT hub areas heavy road traffic is observed and due to work from home employees do not need to face heavy road traffic, and it results in saving precious time. The employees can work with more energy and time. The analysis shows that there is a downside also few employees may not have facilities like the separate room to work, strong Internet connection, they may get disturbance due to family issues. It is also observed that there was more impact on the mental health of employees who are in the beginning phase of their profession. In view of this, it is recommended that there should be some policies in place in order to ensure the good mental health of employees for all cadres, adequate arrangement needs to be deployed by the employer to enable fear-free work culture even in the pandemic period, in turn, it will also improve on efficiency and productivity in long term.

References

1. Bolhari, A., Rezaeean, A., Bolhari, J., Bairamzadeh, S., Soltan, A.A.: The relationship between quality of work life and demographic characteristics of information technology staffs (2011)
2. Igbaria, M., Parasuraman, S., Badawy, M.K.: Work experiences, job involvement, and quality of work life among information systems personnel. MIS Quarterly, 175–201 (1994)
3. Rohini, V., Devi, K., Muthu, N.: Impact of work life balance among software professionals in selected it industries. Eurasian J. Analyt. Chem. **13**(SP), 183–188 (2019)
4. Vijay, S.A., Sekar, P.C.: Work-related quality of life and its association with work productivity among the employees of the Information Technology Enabled Service (ITES) Industries in India. Int. J. Hum. Resour. Manage. **2**(2), 17–26 (2013)
5. Venkatesh, B., Upendra, K.: Women in IT-ITES industries: their coping styles and quality of work life. SDMIMD J. Manage. (2021)

6. Russo, D., Hanel, P.H., Altnickel, S., van Berkel, N.: The daily life of software engineers during the covid-19 pandemic. In: 2021 IEEE/ACM 43rd International Conference on Software Engineering: Software Engineering in Practice (ICSE-SEIP), pp. 364–373. IEEE (2021)
7. Russo, D., Hanel, P.H., Altnickel, S., van Berkel, N.: Predictors of well-being and productivity among software professionals during the COVID-19 pandemic–a longitudinal study. Empir. Softw. Eng. **26**(4), 1–63 (2021)
8. Ford, D., Storey, M.A., Zimmermann, T., Bird, C., Jaffe, S., Maddila, C., Butler, J.L., Houck, B., Nagappan, N.: A tale of two cities: Software developers working from home during the covid-19 pandemic (2020). arXiv preprint arXiv:2008.11147
9. Bhapkar, H.R., Mahalle, P.N., Shinde, G.R., Mahmud, M.: Rough sets in COVID-19 to predict symptomatic cases. In: Santosh, K.C., Joshi, A. (eds.) COVID-19: Prediction, Decision-Making, and Its Impacts. Lecture Notes on Data Engineering and Communications Technologies, vol. 60, pp. 57–68. Springer, Singapore (2021). ISBN 9789811596810
10. Bhapkar, H.R., Mahalle, P.N., Dhotre, P.S.: Virus graph and COVID-19 pandemic: a graph theory approach. In: Hassanien, A.E., Dey, N., Elghamrawy, S. (eds.) Big Data Analytics and Artificial Intelligence Against COVID-19: Innovation Vision and Approach. Studies in Big Data, vol. 78. Springer, Cham (2020). https://doi.org/10.1007/978-3-030-55258-9_2
11. Bhapkar, H.R., Mahalle, P.N., Dey, N., Santosh, K.C.: Revisited COVID-19 mortality and recovery rates: are we missing recovery time period? J. Med. Syst. **44**(12), 202 (2020). https://doi.org/10.1007/s10916-020-01668-6

Chapter 6
Impact of COVID-19 Pandemic on Agriculture Sector

Abstract Agriculture is one of the areas highly affected by lockdown because it depends mainly on labor. The daily work of agriculture requires jobs, and the availability of labor is the main problem. In addition to the other problems caused by the lockdown, i.e., the low supply of pesticides, the reduced opportunity to enter the market and the closure of agricultural equipment stores. It is important to understand how farmers think about the pandemic. Following is an attempt to understand the impact of COVID-19 on agriculture sector using data analysis and mathematical modeling.

6.1 Introduction

Around the globe, India is one of the country where GDP is mainly based on the research as well as agriculture. Being categorized as developing country in India an agriculture is one of the major sources of earning and livelihood. It contributes to almost one-sixth portion to the Indian economy. The entire economy in all the sectors has been affected to large extent, and still there is lot of uncertainty as pandemic is not over yet. In the COVID-19 pandemic lockdown precautionary measures like social distancing and self-isolation have been applied to a slow down rate of infection. As lockdown seems to be very effective strategy for lowering down the number of infections at the same time, it has adverse effects on various businesses and economic fronts [1–3]. Agriculture is one of the domains that have been affected due to lockdown as it is mainly dependent on labors. Day-to-day work in agriculture required labor and due to lockdown availability of laborers was the major issue. Along with this various other problems occurred due to lockdown, i.e., low availability of pesticides, less access to market, shops of agriculture equipment were closed, etc. Various studies have been done to understand the impact of the COVID-19 pandemic on agriculture. It is important to understand the views of farmers regarding this pandemic. In line with this, the main objective of this chapter is to analyze impact of pandemic on work-life balance of farmers through the data resulted from survey. The survey has been carried out in the form of questions.

G. R. Shinde et al., *Quality of Work-Life During Pandemic*, Studies in Big Data 100,
https://doi.org/10.1007/978-981-16-7523-2_6

The main contributions of this study are:

- The preparation of a questionnaire for farmers in their native language for recording their ongoing experiences during this COVID-19 pandemic.
- Preparation of the dataset resulted from the responses collected from the survey initiated for farmers.
- Design a mathematical model to draw statistical inferences, correlations in the form of association rule mining, and key observations.
- Machine learning-based analysis and visualization to draw useful inferences and normative interpretations from the dataset.

6.2 Literature Work

Many researchers have been working to know the impact of COVID-19 on the agriculture sector, the studies are varied as, few are targeted to know the impact on Kharif crop, rabi crop due to availability of labor, availability of pesticides, etc. Arumugam and Kanagavalli [4] presented the challenges that farmers faced during the pandemic and also presented the reaction of the Indian government toward farmers' problems. Varshney et al. [5] presented the impact *of Pradhan Mantri Garib Kalyan Yojana*that is announced by the Indian government immediately after the COVID-19 pandemic, i.e., March 2020. The survey has been targeted to the farmers in Uttar Pradesh, Madhya Pradesh and Rajasthan states. The result of the analysis and survey concluded that the money was transferred to the farmers on time. Gregorioa and Ancog [6] presented the impact of a pandemic on southeast Asian countries. The study says that the 3.11% decrease in aggregate agriculture production and 1.4% decrease in GDP due to pandemic, and this is affected the agricultural food system. Timilsina et al. [7] presented the effect of a pandemic on agriculture and food systems, the study recommends digitization in the agriculture sector to avoid major loss due to pandemics like COVID-19. The need for seeding robots, drones, autonomous tractors, robotic harvesters, quality seeds, easy availability of fertilizers, toll-free numbers for guidance, and flexible financial funding is recommended by this study.

Poudel et al. [8] present a survey of the impact of a pandemic on agriculture at the global level. The food chain and fishery have been affected majorly due to the COVID-19 pandemic. Saha and Bhattacharya [9] present the impact of a pandemic on the agriculture sector in India. This study presents the various challenges farmers faced due to lockdown and few recommendations are also presented in this study to avoid loss due to pandemics in the future. Brewin [10] presented the impact of a pandemic on grain and oilseeds in Canada. The supply chain and market are affected due to social distancing as laborers are required to transport agricultural products from one place to another. Bochtis et al. [11] presented the impact of a pandemic on workers of the agriculture sector. The major reasons for infection and risk factors are presented in this work the various mitigation techniques to reduce loss are also presented in this work. Siche [12] presents the impact of the pandemic of the agriculture sector with respective food demand, food security. The study presents that the food demand slightly decreases due to uncertainty in the COVID-19 pandemic.

Food security is majorly affected due to the lack of availability of laborers. Bhapkar et al. [13] presented mathematical model to reduce symptoms of COVID-19 by using rough set theory approach. Bhapkar et al. [14] defined the first generic formulae to measure the severity of any epidemic or pandemic during its period also, known as the progressive recovery rate and progressive mortality rate.

Many researchers have discussed the impact of the COVID-19 pandemic on the agriculture and food chain sector; however, there is a need to understand the view of farmers regarding the impact of a pandemic. In view of this, the survey of farmer's opinions is conducted and analysis is done. In addition to this, survey indicates that despite having many subsidized schemes and policies for the farmers laid down by government, this pandemic has affected the farmers resulting in the imbalance of the work-life.

The chapter is organized as follows, the description of data and dataset are discussed in Sect. 6.3, statistical description of the dataset is presented in Sect. 6.4. The mathematical modeling for the educational dataset is designed and presented in Sect. 6.5. The data is analyzed and visualized using ML techniques, presented in Sect. 6.6.

6.3 Dataset and Data Description

The survey is carried out for the farmers in Maharashtra State, India. The data is collected during the period, May 2020 to July 2020, 160 farmers have responded to the survey.

The questionnaire consists of a total number of 17questions consist of 2 questions related to knowing demographic information and 15 questions are regarding the availability of human resources, seeds, pesticides, fertilizers, the payment delays for agriculture products, market situation, various packages provided by the government, medical facilities they are getting, how they have handled this pandemic. Initially, it was challenging to collect and prepare data for agriculture sector in the view of reaching to the farmers. The questionnaire was distributed through social media like WhatsApp and in few cases, a Google form link of the questionnaire was provided in person. Questionnaire is shown in Table 6.1.

Table 6.1 Questionnaire

S. No.	Question translation in English
1	What is your age group?
2	How many acres of the field do you have?
3	Are you and your family scared more of getting infected during a pandemic?
4	Are sufficient human resources available for agricultural work during this pandemic?
5	Do you get the required agricultural seeds, pesticides and fertilizers in time?

(continued)

Table 6.1 (continued)

S. No.	Question translation in English
6	Do you get resources or labor for harvesting crops in time?
7	Are you getting proper market prices for your agricultural products during a pandemic?
8	Do you get payment of agricultural product in time?
9	Have you received sufficient support from the government for agricultural development during the pandemic?
10	Are you satisfied with the protective and preventive measures taken by the government?
11	Due to the supply chain slowdown in a pandemic, is it affecting your economy?
12	Do you feel that you have handled the situation of pandemics very well?
13	Is the agricultural sector a bright spot in lockdown gloom?
14	Are you agree that the government should promote trade by avoiding export bans and import restrictions to boost the agricultural sector?
15	Are you getting desired medical facilities in your village or town during a pandemic?
16	How much satisfactory justice is given to the health and education of your family?
17	Rate the level of mental disturbance during the period of a pandemic?

6.4 Descriptive Statistics

Dataset can be described briefly with statistical information. Statistical information includes a measure of central tendency and a measure of data dispersion. The standard deviation is the measure of data dispersion, and the mean, mode, median are the measure of the central tendency of the dataset. Based on the mean, median and mode values symmetry of data can be understood. The central tendency of the dataset describes: How data is distributed? How it is spread from its mean value? What is the answer that maximum times appeared in responses of participants? A study is also carried on this underline dataset to analyze the distribution in order to summarize nature of the dataset, variability as well as association between the categorical fields.

The qualitative statistical description of this dataset is shown in Table 6.2. The Pandas library in Python is used to calculate statistical values of each, the features are renamed as U1 to U17, and in further sections of the chapter, these naming conventions are used.

Table 6.2 Statistical values

Demographic/variables	Std deviation	Mean	Mode	Median	Max	Min
U_1: Age	1.095858	2.41875	2	2	5	1
U_2: Acres field	1.329338	2.2375	1	2	5	1
U_3:Scared of Infection	1.186335	2.5375	3	3	5	1
U_4:Availability of human resources	1.222802	3.18125	2	3	5	1

(continued)

Table 6.2 (continued)

Demographic/variables	Std deviation	Mean	Mode	Median	Max	Min
U_5: Availability of seeds, pesticides and fertilizers	1.216226	2.89375	2	3	5	1
U_6: Availability of resources or labors	1.1152	2.81875	2	3	5	1
U_7: Market prices	0.797401	1.675	1	1	4	1
U_8: Payment of the agricultural product	0.954219	2.2875	2	2	5	1
U_9: Government support	1.103169	1.875	1	1	5	1
U_{10}: Protective and preventive measures taken by the government	1.303584	2.60625	3	3	5	1
U_{11}: Effect of supply chain slowdown	0.939168	4.13125	5	4	5	1
U_{12}: Handled situation of pandemic	1.086492	3.45625	3	4	5	1
U_{13}: Agricultural sector a bright spot	1.105856	4.23125	5	5	5	1
U_{14}:Ban on import and export	1.378619	3.79375	5	4	5	1
U_{15}: Medical facilities	1.151138	2.35625	2	2	5	1
U_{16}: Health and education of your family	1.187444	2.39375	3	3	5	1
U_{17}: Mental disturbance	1.163856	3.4375	4	4	5	1

6.5 Mathematical Modeling

There were 17 questions. We had received 160 responses, those have all the required data. In this question set, there were 2 general questions and 15 questions related to the effect of COVID-19 on the work-life of farmers.

To illustrate the rough analysis, with the above-received data. We would also like to verify the effect of COVID-19 on the mental health of an individual in the agricultural sector by considering the above dataset. We have selected seven questions and 16 responses from the dataset as a sample case. The mental health of people is described by using the seven criteria $U_3, U_5, U_7, U_{10}, U_{11}, U_{13}$ and U_{15} together with the corresponding scales ordered from 0 to 4, given below.

U_3: Scared of getting infected: $\{1, 2, 3, 4, 5\}$
U_5: Get required agricultural materials in time: $\{1, 2, 3, 4, 5\}$
U_7: Proper market price: $\{1, 2, 3, 4, 5\}$
U_{10}: Government measures: $\{1, 2, 3, 4, 5\}$
U_{11}: Effect on personal economy: $\{1, 2, 3, 4, 5\}$
U_{13}: Agricultural sector bright spot: $\{1, 2, 3, 4, 5\}$
U_{15}: Desired facilities at village: $\{1, 2, 3, 4, 5\}$.

Therefore, the set of condition attributes is $C = \{U_3, U_5, U_7, U_{10}, U_{11}, U_{13}, U_{15}\}$. Sixteen responses with rather different responses have been sorted by us after due to mathematical consideration. The set of decision attributes is.

$D = \{\text{Yes, No}\}$, where Yes: COVID-19 is affected on the mental disturbance during pandemic period. No: COVID-19 is not affected on the mental disturbance during pandemic period. The desired information is presented in Table 6.3.

From the table, we have,

$X_Y = \{R_1, R_2, R_3, R_6, R_7, R_9, R_{11}, R_{12}, R_{15}, R_{16}\}$ and $X_N = \{R_4, R_5, R_8, R_{10}, R_{13}, R_{14}\}$.

The lower approximation and upper approximation of the rough set of X_Y and X_N are given below.

$\underline{CX_Y} = \{R_1, R_2, R_3, R_6, R_7, R_9, R_{11}, R_{15}, R_{16}\} = X_Y - \{R_{12}\}$.

$\overline{CX_Y} = \{R_1, R_2, R_3, R_6, R_7, R_9, R_{11}, R_{12}, R_{14}, R_{15}, R_{16}\} = X_Y \text{ U } \{R_{14}\}$ and.

$\underline{CX_N} = \{R_4, R_5, R_8, R_{10}, R_{13}\} = X_N - \{R_{14}\}$.

$\overline{CX_N} = \{R_4, R_5, R_8, R_{10}, R_{12}, R_{13}, R_{14}\} = X_N \text{ U } \{R_{12}\}$.

Moreover, $\text{Bnd}_C (X_Y) = \{R_{12}, R_{14}\}$ and $\text{Bnd}_C (X_N) = \{R_{12}, R_{14}\}$.

The next step of the rough set analysis of the decision table is the construction of a minimal subset of independent criteria ensuring the same quality of sorting as whole set C, that means reduces of *C*. The following are the required reduces in our case.

$RED^1{}_X = \{U_3, U_5, U_{11}\}$, $RED^2{}_X = \{U_3, U_{10}, U_{11}\}$, $RED^3{}_X = \{U_5, U_{10}, U_{11}, U_{13}, U_{15,}\}$

And $RED^4{}_X = \{U_7, U_{10}, U_{13}\}$, $RED^5{}_X = \{U_5, U_{11}, U_{13}\}$,

It is observed that the criteria U_7 is not present in the positive decision.

Table 6.3 Decision table for agricultural sector

Condition/ Responses	U_3	U_5	U_7	U_{10}	U_{11}	U_{13}	U_{15}	Decision
R_1	1	2	2	2	4	3	2	Yes
R_2	1	1	3	1	5	2	3	Yes
R_3	1	2	1	1	5	5	1	Yes
R_4	1	2	1	5	3	4	5	No
R_5	2	2	1	3	4	5	3	No
R_6	1	2	3	3	4	1	1	Yes
R_7	3	3	2	3	3	3	2	Yes
R_8	1	5	2	1	4	4	3	No
R_9	5	2	4	1	5	5	1	Yes
R_{10}	3	2	2	4	3	4	1	No
R_{11}	2	3	2	2	4	5	2	Yes
R_{12}	1	3	3	3	5	5	5	Yes
R_{13}	2	5	3	3	3	5	2	No
R_{14}	1	3	3	3	5	5	5	No
R_{15}	3	2	3	5	5	5	4	Yes
R_{16}	5	5	4	5	3	1	4	Yes

Case 1: Select $RED^1{}_X$

Suppose we have chosen to reduce $RED^1{}_X$ composed of U_3, U_5, U_{11}, that is, scared of getting infected, get required agricultural materials in time and effect on the personal economy. Now the decision table can be reduced to criteria given in $RED^1{}_X$ which is given in Table 6.4.

The decision rules or laws are generated from the reduced decision table that has the following forms as shown in Table 6.5.

Now, consider Table 6.6 of certainty factor and coverage factor of the above rules.

By the certainty factors of the decision rules, we can conclude the following.

Table 6.4 Reduced decision table for $RED^1{}_X$

Condition/responses	U_3	U_5	U_{11}	Decision
R_1	1	2	4	Yes
R_2	1	1	5	Yes
R_3	1	2	5	Yes
R_6	1	2	4	Yes
R_7	3	3	3	Yes
R_9	5	2	5	Yes
R_{11}	2	3	4	Yes
R_{12}	1	3	5	Yes
R_{15}	3	2	5	Yes
R_{16}	5	5	3	Yes
R_4	1	2	3	No
R_5	2	2	4	No
R_8	1	5	4	No
R_{10}	3	2	3	No
R_{13}	2	5	3	No
R_{14}	1	3	5	No

Table 6.5 Decision Rules for $RED^1{}_X$

Rule No. L_i	Rule	Decision (d, Y/N)
1	(U_3, 5)	Y
2	(U_5, 1) ^ (U_{11}, 5)	Y
3	(U_3, 1) ^ (U_5, 2) ^ (U_{11}, 4)	Y
4	(U_3, 3) ^ (U_5, 3) ^ (U_{11}, 3)	Y
5	(U_3, 1) ^ (U_5, 3) ^ (U_{11}, 5)	Y
6	(U_3, 1) ^ (U_5, 3) ^ (U_{11}, 5)	N
7	(U_3, a) ^ (U_5, b) ^ (U_{11}, c) and $a + b + c > 10$	Y

Table 6.6 Certainty and coverage factor

Rule	Certainty factor	Coverage factor
1	1	(2/10) = 0.2
2	1	0.1
3	1	0.2
4	1	0.1
5	0.5	0.1
6	0.5	0.16
7	1	0.2

- If the score of U_3 is 5, then COVID-19 is affected on the mental health certainly during the pandemic period. That is, if someone is scared more of getting infected during a pandemic, then COVID-19 is affected on the mental health certainly during the pandemic period.
- If the score of U_5 is 1 and the score of U_{11} is 5, then COVID-19 is affected on the mental health certainly during the pandemic period.
- If the score of U_3 is 1, U_5 is 2 and U_{11} is 4, then COVID-19 is affected on the mental health certainly during the pandemic period.
- If the score of U_3, U_5 and U_{11} is 3, then COVID-19 is affected on the mental health certainly during the pandemic period.
- If the score of U_3, U_5 and U_{11} is 1, 3, 5, respectively, then the possibility that COVID-19 is affected mental health during the pandemic period is 50%.
- If the score of U_3, U_5 and U_{11} is 1, 3, 5, respectively, then the possibility that COVID-19 is not affected on the mental health during the pandemic period is 50%.
- If the score of U_3, U_5 and U_{11} is *a*, *b*, *c*, respectively, and $a + b + c > 10$ then the COVID-19 is affected certainly on the mental health during the pandemic period.

Case 2: Select $RED^2{}_X$

Suppose we have chosen to reduce $RED^1{}_X$ composed of U_3, $U_{10,}$ U_{11}, that is, scared of getting infected, get required agricultural materials in time and effect on the personal economy. Now the decision table can be reduced to criteria given in $RED^2{}_X$ which is given in Table 6.7.

The decision rules or laws are generated from the reduced decision table that has the forms as shown in Table 6.8.

Now, consider Table 6.9 for certainty factor and coverage factor of the above rules. By the certainty factors of the decision rules, we can conclude the following.

- If the score of U_3 is 5, then COVID-19 is affected on the mental health certainly during the pandemic period.
- If the score of U_{10} is 4, then COVID-19 is not affected on the mental health certainly during the pandemic period.
- If the score of U_3 is 1, U_{10} is 2 and U_{11} is 5, then COVID-19 is affected on the mental health certainly during the pandemic period.

Table 6.7 Decision table for $RED^2{}_X$

Condition/responses	U_3	U_{10}	U_{11}	Decision
R_1	1	2	4	Yes
R_2	1	1	5	Yes
R_3	1	1	5	Yes
R_6	1	3	4	Yes
R_7	3	3	3	Yes
R_9	5	1	5	Yes
R_{11}	2	2	4	Yes
R_{12}	1	3	5	Yes
R_{15}	3	5	5	Yes
R_{16}	5	5	3	Yes
R_4	1	5	3	No
R_5	2	3	4	No
R_8	1	1	4	No
R_{10}	3	4	3	No
R_{13}	2	3	3	No
R_{14}	1	3	5	No

Table 6.8 Decision rules $RED^2{}_X$

Rule No. L_i	Rule	Decision (d, Y/N)
1	$(U_3, 5)$	Y
2	$(U_{10}, 4)$	N
3	$(U_3, 1)$ ^ $(U_{10}, 1)$ ^ $(U_{11}, 5)$	Y
4	$(U_3, 3)$ ^ $(U_{10}, 3)$ ^ $(U_{11}, 3)$	Y
5	(U_3, a) ^ (U_{10}, b) ^ (U_{11}, c) and $a + b + c = 13$	Y

Table 6.9 Certainty and coverage factor $RED^2{}_X$

Rule	Certainty factor	Coverage factor
1	1	(2/10) = 0.2
2	1	0.166
3	1	0.2
4	1	0.1
5	1	0.2

- If the score of U_3, U_5 and U_{11} is 3, then COVID-19 is affected on the mental health certainly during the pandemic period.
- If the score of U_3, U_{10} and U_{11} is a, b, c, respectively, and $a + b + c = 13$, then the COVID-19 is affected certainly on the mental health during the pandemic period.

Similarly, we can apply for the remaining reduces 3, 4 and 5.

6.6 Machine Learning-Based Analysis and Visualization

The responses of survey data are recorded in an excel sheet, and these are in textual form. The encoding is done to convert textual data into the numerical format. The machine learning approach is used for data visualization and analysis. The main objective of applying ML on this dataset is to carry out the behavioral as well as the normative analysis. In this study, the k-mean algorithm is used for clustering. The analysis is done to understand the impact of COVID-19 on the availability of human resources, labors, seeds, pesticides and market price. Figure 6.1(i) shows the graphs of the impact of the COVID-19 pandemic on the availability of resources with respect to the acres of area own by the farmer. The graph shows that there was a lack of resources available irrespective of how much area farmers own. Figure 6.1(i) and (ii) shows the graphs of the impact of COVID-19 on the availability of human resources (laborers) with respect to the acres of area own by the farmer. Figure 6.1(iii) and (iv) shows the graphs of the impact of COVID-19 on the availability of seeds, pesticides and fertilizers with respect to the acres of area own by the farmer.

The graph shows that the responses of farmers are in favor of lack of availability of resources, seeds, pesticides and fertilizers. The farmers who belong to the large field are impacted less than farmers who have fewer acres of the farm. To summarize the analysis shows that there was labor issue faced by all the farmers during this pandemic and same issue is likely to increase if this pandemic will continue. In the view of this, there should be proper policies laid down by authorities in the favor

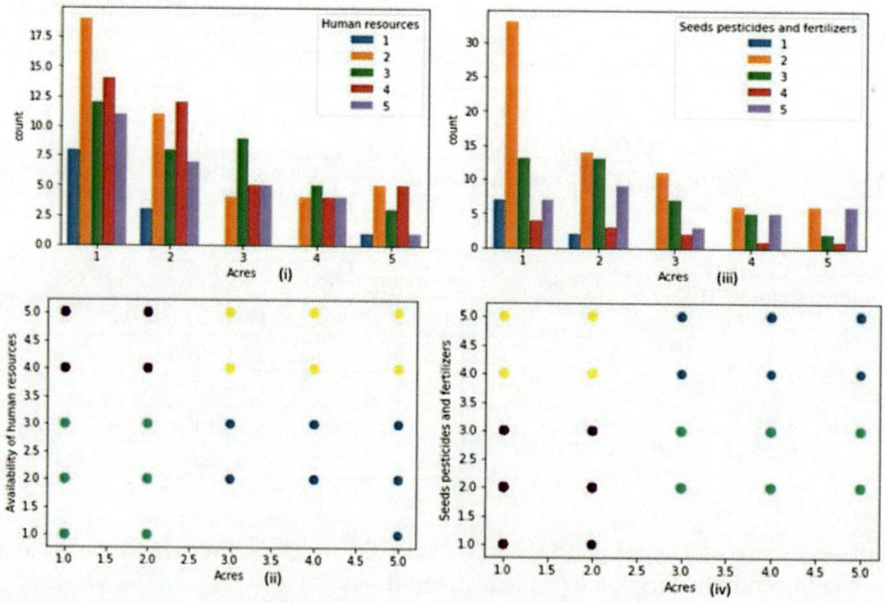

Fig. 6.1 Analysis of COVID-19 on availability of human resources, seeds, pesticides and fertilizers

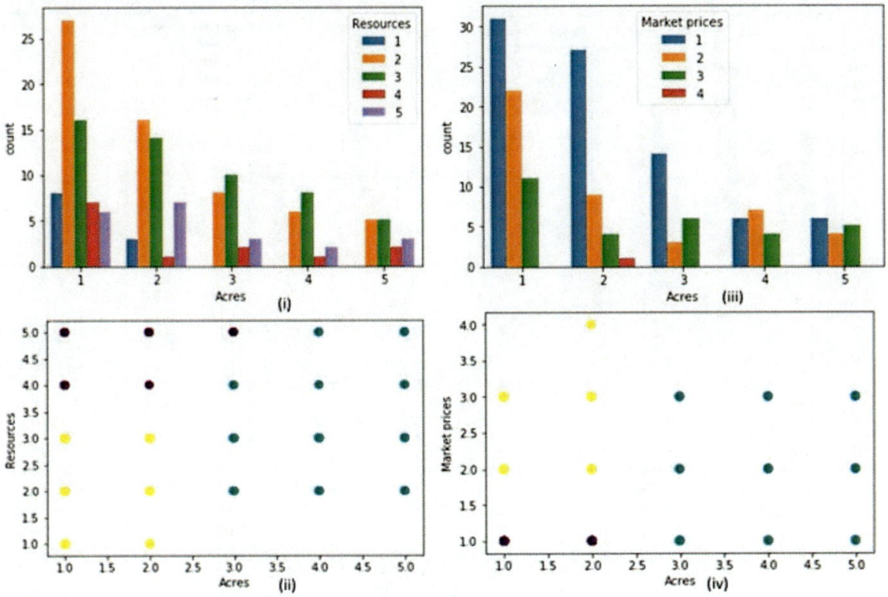

Fig. 6.2 Analysis of COVID-19 on resources and market price

of labors with respect to the wages, terms/conditions for the payment of wages and assurance of the employability.

Figure 6.2(i) and (ii) show the graphs of the impact of COVID-19 on the availability of resources (equipment) with respect to the acres of area own by the farmer. Figure 6.2(iii) and (iv) shows the graphs of the impact of COVID-19 on the market price that farmers are getting for their agriculture products with respect to the acres of area own by the farmer.

The graph shows that the responses of farmers are in favor of lack of availability of resources and less price they are getting for agriculture goods. Regarding the availability of resources, farmers who belong to the large field are impacted less than farmers who have fewer acres of the farm. Due to this pandemic, the allied market for agricultural sector is highly impacted resulting in the imbalance in the prices of goods and equipments. The demand and supply is also disturbed due to this uncertain situation of COVID-19. ML-based analysis also shows that farmers received considerably low compensation and appreciation for their efforts. The result shows that there is major cut down in the cost they have received from the sale of agriculture products. The interesting observations from this analysis reveal that farmers with more acres of field are impacted more than famers with less farm. The formal analysis also shows that this cut down in the cost is likely to continuing with pandemic. The policies are required with respect to the need-based distribution across geographical areas at local, national and international levels. Also, there should be proper regulatory support in order operationalize these policies for the farmers by the government.

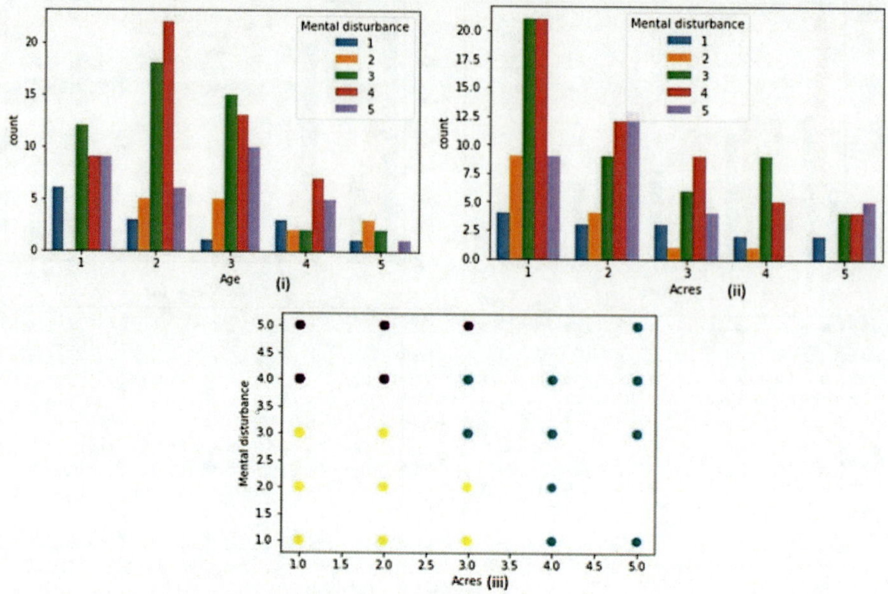

Fig. 6.3 Analysis of COVID-19 on mental health of farmers

The mental health is one of the important qualitative parameters which is affected more in the farmers due to this pandemic. Figure 6.3(i) shows the graph of the impact of COVID-19 on mental health with respect to the age of the farmer. The graph shows that middle-aged farmers are impacted more than others. The reason for this impact is the prospective planning and investment made by middle-aged farmers on modernization of farming methodology and extension planning.

Figure 6.3(ii) and (iii) shows the graph of the impact of COVID-19 on mental health with respect to the acres of area own by the farmer. The impact of mental health is found to be varied among farmers having different acres of land. The graph shows that farmers who belong to the large field are impacted less than farmers who have fewer acres of the farm.

6.7 Summary

Being India as one of the agriculture-based countries, national economy is relying on revenue generated from this sector. This chapter has analyzed the impact of COVID-19 pandemic on the work-life balance of the farmers. Mathematical modeling and machine learning-based analysis are presented and discussed. The analysis shows that in totality all the farmers are highly impacted with respect to human resources, equipment, seeds, pesticides, fertilizers and market price. It is also seen that the farmers with less acres of field are impacted more for all the parameters as compared to famers who belongs to higher acres of farm. The rough sets are found to be more appropriate tool of mathematics for model building of the agriculture dataset. The mathematical modeling based on the rough set theory is discussed in the scope of this chapter.

References

1. Sahoo, P., Ashwani: COVID-19 and Indian economy: Impact on growth, manufacturing, trade, and MSME sector. Glob. Bus. Rev. **21**(5), 1159–1183 (2020)
2. McKibbin, W., Fernando, R.: The economic impact of COVID-19. Econ Time COVID-19 **45**(10.1162) (2020)
3. He, H., Harris, L.: The impact of Covid-19 pandemic on corporate social responsibility and marketing philosophy. J. Bus. Res. **116**, 176–182 (2020)
4. Arumugam, D.U., Kanagavalli, D.G.: COVID-19: impact of agriculture in India. Aegaeum. J. **8**(5) (2020)
5. Varshney, D., Kumar, A., Mishra, A.K., Rashid, S., Joshi, P.K.: India's COVID-19 social assistance package and its impact on the agriculture sector. Agric. Syst. **189**, 103049 (2021)
6. Gregorioa, G.B., Ancog, R.C.: Assessing the impact of the covid-19 pandemic on agricultural production in Southeast Asia: toward transformative change in agricultural food systems. Asian J. Agric. Dev. **17**(1362-2020-1097), 1–13 (2020)
7. Timilsina, B., Adhikari, N., Kafle, S., Paudel, S., Poudel, S., Gautam, D.: Addressing the impact of COVID-19 post pandemic on farming and agricultural deeds. Asian J. Adv. Res. Rep., 28–35 (2020)
8. Poudel, P.B., Poudel, M.R., Gautam, A., Phuyal, S., Tiwari, C.K., Bashyal, N., Bashyal, S.: COVID-19 and its global impact on food and agriculture. J. Biol. Today's World **9**(5), 221–225 (2020)
9. Saha, T., Bhattacharya, S.: Consequence of lockdown amid COVID-19 pandemic on Indian agriculture. Food Sci. Rep. **1**(special issue), 47–50 (2020)
10. Brewin, D.G.: The impact of COVID-19 on the grains and oilseeds sector. Can. J. Agric. Econ./Revue canadienne d'agroeconomie **68**(2), 185–188 (2020)
11. Bochtis, D., Benos, L., Lampridi, M., Marinoudi, V., Pearson, S., Sørensen, C.G.: Agricultural workforce crisis in light of the COVID-19 pandemic. Sustainability **12**(19), 8212 (2020)
12. Siche, R.: What is the impact of COVID-19 disease on agriculture? Scientia Agropecuaria **11**(1), 3–6 (2020)
13. Bhapkar, H.R., Mahalle, P.N., Shinde, G.R., Mahmud, M.: Rough sets in COVID-19 to predict symptomatic cases. In: Santosh, K.C., Joshi, A. (eds.) COVID-19: Prediction, Decision-Making, and Its Impacts. Lecture Notes on Data Engineering and Communications Technologies, vol. 60, pp. 57–68. Springer, Singapore (2021). ISBN 9789811596810
14. Bhapkar, H.R., Mahalle, P.N., Dey, N., Santosh, K.C.: Revisited COVID-19 mortality and recovery rates: are we missing recovery time period? J. Med. Syst. **44**(12), 202 (2020). https://doi.org/10.1007/s10916-020-01668-6

Chapter 7
Exploratory Data Analysis

Abstract Data analysis is the process of extracting insights from data. Data is heterogeneous in all ways, and processing such data is a challenge. Before applying any machine learning model to any datasets, it is necessary to understand the problem, deal with the missing values and noise, visualize the dataset, and to select the machine learning model to analyze the data. Following is an attempt to highlight the techniques of data analysis.

7.1 Principle Component Analysis Overview

Data analysis is the process to take out insights from the data. Data can be in any form that it may be structured or unstructured, it may be of any size big or small, it may consist of any type of data, i.e., real, dynamic or static, it may be retrieved from any source, i.e., Internet, organizations. Concisely, we can say that the data is heterogeneous in all the forms, handling such data is challenging. There is a need to follow step-by-step process before applying machine learning model on dataset [1], i.e., understand the problem, handle the missing values and noise, visualize the dataset, and the last step is to choose the machine learning model that can be applied on a dataset for analysis. While selecting the ML model, there is also a need to think about features of the dataset that are impacting more on the outcome of the analysis.

In many applications, the dataset may have a large number of variables/features then handling, visualizing, and analyzing such dataset is complex. Higher dimension results in a larger volume of space for data, and hence, data samples become sparse. In the statistical approach, finding out similar properties between sample points which are distributed in larger volumes is difficult, and sparsity increases exponentially with the dimensionality of the dataset. Hence, there is a need to convert such a dataset to a simpler one by selecting only a few features for analysis and here comes the role of dimensionality reduction. The principal component analysis (PCA) is used to reduce the dimensionality of the dataset [1, 2]. PCA supports to decide that among the all features which feature should be selected for analysis. While converting the dataset, there is a need to find out which features needed to drop from analysis and which needed to select for the analysis. The features that make a high impact on

G. R. Shinde et al., *Quality of Work-Life During Pandemic*, Studies in Big Data 100,
https://doi.org/10.1007/978-981-16-7523-2_7

the outcome are selected for the analysis. It may possible that few features having a similar impact (correlated features) on the outcome, then among those features only one feature can be selected for analysis. The dimensionality reduction method converts datasets with a large number of features into a dataset that consists of a small number of combination of uncorrelated features, i.e., principal components. The number of features is reduced, and hence, data loss will be there; however, in PCA it is done with minimal data loss. In this, focus is kept to take out maximum information from an original dataset with few features, i.e., less data loss.

7.2 PCA for the Categorical Survey Dataset

In various applications, datasets are generated through surveys, in survey participant's opinions are taken in the form of questions. Each question is supported with some predefined answers, and participants are selecting one of the answers from the list of answers provided. The dataset generated from such survey is of the categorical form, generally both types of the variables/features; i.e., ordinal and nominal are present in such types of datasets. In the categorical dataset, it is important to know the similarities between the features. The computation time and complexities can be minimized by forming a group of similar features and selecting one of the features from the group for the analysis, i.e., dimensionality reduction. PCA can be applied to select features that are linearly uncorrelated to each other, and these can provide the maximum information which is provided by of complete dataset, i.e., principal component[3]. In the literature, PCA is applied to categorical data for dimensionality reduction[4–8].

In Chap. 4, the educational dataset has been analyzed, dataset contains 17 features/variables, i.e., U1: Age, U2: Qualification, U3: Gender, U4: Job Profile, U5: Experience, U6: Mental health, U7: Security by Company, U8: Job security, U9: Teaching, U10: A pay cut, U11: Impact, U12: Arrangements, U13: Career shift, U14: Family crisis, U15: Skill sets improvement, U16: Work-life balance, U17: Learning performance. The PCA has been applied, and based on linearly uncorrelated PCA features are picked up for further analysis, i.e., U6: Mental health, U7: Security by Company, U10: A pay cut, U12: Arrangements, U13: Career shift, U16: Work-life balance.

7.3 Case Studies

In the literature, researchers have used PCA for dimensionality reduction [3–7]. In various applications, the number of features/variables is large in number, and hence to reduce the computational complexities, it is recommended to drop features that are not important. In such cases, PCA can be applied, and in line with this, understanding the working of PCA is important. The steps involved in the computation of PCA and interpretation of PCA can be understood more clearly by discussing the case study. Let us take the dataset that is generated from the survey of the educational sector as a case study. A detailed discussion of the dataset and analysis is presented in Chap. 4. Total 17 variables/features are there in the dataset, and PCA is applied on this dataset for dimensionality reduction. After the application of PCA, six features are selected for the analysis. Let us discuss the computation of principal components for an educational dataset in the below section.

7.3.1 PCA Computations

PCA converts dataset with a large number of correlated/uncorrelated features into the dataset with less number of linearly uncorrelated features. The steps for computing principal components are as follows:

Step 1: Standardization of the values in the dataset

The various analysis algorithms can work efficiently on the dataset in which all feature values are on the same scale. There are various reasons for which standardization of dataset is required.

Observations in the dataset may be measured in different units then algorithms that work on the distance, i.e., clustering algorithms, SVM, etc., will lead to wrong results. The feature which has a higher range will be dominating than other features, e.g., in the application where analysis needed to be done on the features height and weight. The height is measured in meters [1–3], and weight is measured in pounds, i.e., [20–200]; in such situations, the weight feature will dominating in any distance-based machine learning algorithm. There is another reason for to standardization of the dataset is the wrong selection of principal components while applying PCA. When PCA is applied on a dataset that is not on the same scale, then the feature with a broad range will be dominating in the selection of principal components as it has values in a broad range. The selection of principal components can be lead to the wrong selection, to avoid this dataset values should be on the same scale.

In standardization, all the values will have a mean of zero and unit standard deviation. The simplest method that is used to standardize values is Z-score. In the Z-score, each observation value is subtracted by its mean value and divided by the standard deviation.

Step 2: Computing the covariance matrix

In the PCA, aim is to find the set of features that have high variance and these are highly uncorrelated to each other. The covariance shows a correlation between two or more features, the positive correlation means features values are increasing/decreasing with each other in the same direction. A negative correlation means features values are increasing/decreasing in opposite direction. The covariance matrix represents the variance between the features, the dataset is of 'n' dimension, the covariance matrix is of '$n \times$ n' dimension.

Step 3: Computing the eigenvalues and eigenvectors

The eigenvectors are the principal components of the covariance matrix; i.e., it is representing the direction of the maximum covariance and its magnitude is the eigenvalue. The eigenvalues and eigenvectors are calculated, and a matrix of eigenvectors is formed.

Step 4: Sorting of the eigenvalues and their corresponding eigenvectors

The eigenvalues are sorted with a magnitude of eigenvalues. The sorting is done as in PCA, features that have high variance are selected as principal components.

Step 5: Select *k*, i.e., number of principal components

In this step, the dimensionality of the dataset is determined, i.e., which features and how many features should be taken ahead for analysis is determined. To select the '*k*' value, the plot of cumulative addition of eigenvalues is plotted and from that plot, '*k*' is decided. Suppose the due to the first 2 principal components 94% of the variance is achieved, then these two components are selected.

Step 6: Transform the dataset in newly computed *k*-dimension

This is the last step of PCA, the '*n*' dimension dataset is converted into the '*k*'-dimension dataset. The original dataset with '*n*' dimension will be converted into a dataset with k features which will provide maximum information that can be taken from an original dataset with '*n*' dimensions. This can be done by multiplying the transpose of the original dataset by transpose and the dataset is projected with this *k*-dimensions for further analysis.

7.3.2 Interpreting the PCA

The PCA is used for dimensionality reduction, and it is also used for taking out meaningful information from the dataset. The first step is to understand the contribution of principal components toward the total variance. The second step is to understand the contribution of each feature in specific principal components. The bar plots can be plotted to know how much each component contributing to the total variance. The first few principal components may contribute a maximum of variance then these components are considered for further analysis.

The second step is to find the features that are contributing more to the specific principal component. The factor loading can be plotted to know the behavior of features, i.e., features that are close to each other. The positive loading shows that the particular feature is positively correlated to the principal component. The negative loading shows that the feature is negatively correlated to the principal component. The computation can be done to find out the average value of loading considering that all the features were contributing the same portion to the principal component. The features that having loading larger than this value are considered important features for analysis as these are contributing more toward outcomes.

The python language and sklearn library are used to compute principal components of the educational sector database. The contribution of each component in total variance can find out by plotting the graph of principal components versus variance, as shown in Fig. 7.1

The two principal components achieve 42.31% of the variance, and 8 components achieve 86.59% of the variance, as shown in Figs. 7.2 and 7.3, respectively.

Once the number of principal components is finalized based on total variance achieved, the next step to finding features that have the highest loading can be done by checking each features contribution in the respective principal component, as shown in Table 7.1

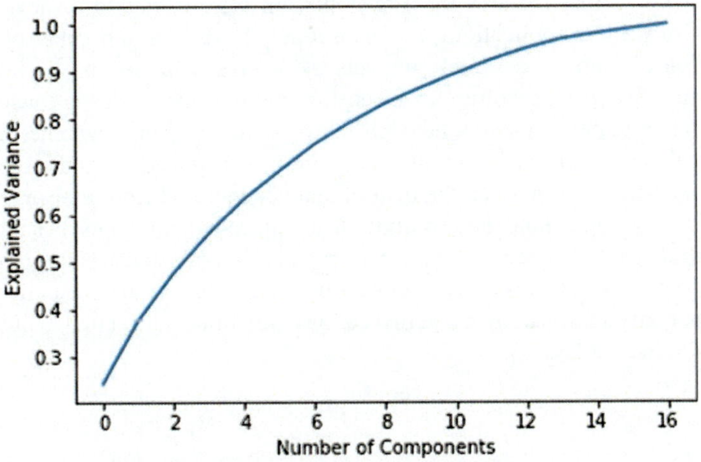

Fig. 7.1 Number of principal components versus explained variance

```
from sklearn.decomposition import PCA
pca= PCA(n_components=2)
pca.fit(df[['Mental health', 'Security by Company', 'Job security', 'Teaching',
        'Pay cut', 'Impact', 'Arrangements', 'Career shift', 'Family crisis ',
        'Skill sets improvement', 'Work-life balance', 'learning performance']])
pca_data=pca.transform(df[['Mental health', 'Security by Company', 'Job security', 'Teaching',
        'Pay cut', 'Impact', 'Arrangements', 'Career shift', 'Family crisis ',
        'Skill sets improvement', 'Work-life balance', 'learning performance']])
total_var = pca.explained_variance_ratio_.sum() * 100
print('Total Explained Variance: ',total_var)

Total Explained Variance:  42.317052347718345
```

Fig. 7.2 Total explained variance: two principal components

```
from sklearn.decomposition import PCA
pca= PCA(n_components=8)
pca.fit(df[['Mental health', 'Security by Company', 'Job security', 'Teaching',
        'Pay cut', 'Impact', 'Arrangements', 'Career shift', 'Family crisis ',
        'Skill sets improvement', 'Work-life balance', 'learning performance']])
pca_data=pca.transform(df[['Mental health', 'Security by Company', 'Job security', 'Teaching',
        'Pay cut', 'Impact', 'Arrangements', 'Career shift', 'Family crisis ',
        'Skill sets improvement', 'Work-life balance', 'learning performance']])
total_var = pca.explained_variance_ratio_.sum() * 100
print('Total Explained Variance: ',total_var)

Total Explained Variance:   86.59347375389592
```

Fig. 7.3 Total explained variance: eight principal components

7.4 Summary

It is important to find features that are important for analysis to achieve accurate outcomes. In various applications, the dataset may have a large number of features, the selection of features definitely impacts the analysis. In various applications, it may possible that if all features are taken for analysis, then it leads toward wrong analysis as few features may not be contributing to the outcome. Another reason for dimensionality reduction is handling, analyzing and visualizing, and larger dimension data is difficult. Moreover, the dataset that is generated from the survey is very much difficult to understand the contribution of features on outcomes. In such types of datasets, dimensionality reduction plays a crucial role as these datasets contain categorical information. PCA is used to reduce the dimension of the dataset by selecting features that have high variance and that can provide information that is provided by all the features together.

Table 7.1 Component loadings

Variable	PC1	PC2	PC3	PC4	PC5	PC6
Mental health	−0.215692	−0.3071814	−0.2350573	−0.4305631	0.33783322	−0.1247046
Security by company	0.37599952	0.20260795	−0.3217978	−0.1608391	0.07017463	−0.2448966
Job security	−0.330644	−0.2448047	−0.2845571	−0.0844466	0.19860776	−0.3266175
Teaching	0.24353296	−0.4431321	0.11117579	0.09829489	0.24243095	0.39231431
Pay cut	0.36110821	0.16505841	−0.3069718	−0.0278669	0.15382948	−0.0629013
Impact	−0.0678004	0.00030138	−0.718144	0.40055229	−0.0988238	0.48391297
Arrangements	0.41678489	0.128449	−0.2149677	−0.1912664	0.17106377	−0.1589911
Career shift	−0.3206381	−0.2513717	−0.2308146	0.24019854	−0.0284801	−0.3353364
Family crisis	0.05483517	−0.2620209	−0.1746445	−0.5637579	−0.7233323	0.14773203
Skill sets improvement	0.2596405	−0.4427975	0.02625898	0.25269676	−0.1604253	−0.1699199
Work-life balance	0.31143296	−0.2261322	0.03044714	0.3444487	−0.276421	−0.4448272
learning performance	0.2540522	−0.4307001	0.07934285	−0.1270145	0.30637959	0.19864322
Variable	PC7	PC8	PC9	PC10	PC11	PC12
Mental health	−0.6036813	0.15817511	0.14841539	0.12021583	0.22015561	−0.1199981
Security by company	0.12513763	0.12266608	−0.4564968	−0.1843358	0.0912509	−0.5846797
Job security	0.30503106	−0.160989	0.1150124	−0.5408155	−0.4041687	0.09216429

(continued)

Table 7.1 (continued)

Variable	PC7	PC8	PC9	PC10	PC11	PC12
Teaching	0.31001756	0.31398625	0.23570125	−0.3331746	0.36927765	−0.1095252
Pay cut	0.17115911	0.32297739	0.53866194	0.37597385	−0.3929197	0.01481573
Impact	−0.2026774	−0.1700586	−0.0507425	−0.0513304	0.00773717	0.00497041
Arrangements	0.03444993	−0.2081872	−0.0850388	−0.1032852	0.36542941	0.69028218
Career shift	0.43623784	0.12272656	−0.1460279	0.49412567	0.37058278	0.03987349
Family crisis	0.16902716	−0.0011665	0.074772	0.02073097	0.00563973	0.01887339
Skill sets improvement	−0.2786638	0.44219376	−0.3885541	−0.038874	−0.3387576	0.27736607
Work-life balance	−0.2208707	−0.3807562	0.41461936	−0.0846341	0.17543822	−0.233264
learning performance	0.11782701	−0.5507076	−0.2259182	0.37437776	−0.2694502	−0.1150487

References

1. Karamizadeh, S., Abdullah, S.M., Manaf, A.A., Zamani, M., Hooman, A.: An overview of principal component analysis. J. Signal Inf. Process. **4**(3B), 173 (2013)
2. Abdi, H., Williams, L.J.: Principal component analysis. Wiley Interdiscip. Rev. Comput. Stat. **2**(4), 433–459 (2010)
3. Online available at https://rpubs.com/Shreyansh_Shivam/604562. Accessed: 30 July 2021
4. Šulc, Z., Řezanková, H.: Dimensionality reduction of categorical data: comparison of HCA and CATPCA approaches. In: Proceedings of the 18th International Scientific Conference Applications of Mathematics and Statistics in Economics (AMSE), Jindrichuv Hradec, Czech Republic, pp. 2–6 (2015)
5. de Sousa Mendes, G.H., Miller Devos Ganga, G.: Predicting success in product development: the application of principal component analysis to categorical data and binomial logistic regression. J. Technol. Manag. Innov. **8**(3), 83–97 (2013)
6. Niitsuma, H., Okada, T.: Covariance and PCA for categorical variables. In: Pacific-Asia Conference on Knowledge Discovery and Data Mining, pp. 523–528. Springer, Berlin, Heidelberg (2005)
7. Chavent, M., Kuentz, V., Saracco, J.: A partitioning method for the clustering of categorical variables. In: Classification as a Tool for Research, pp. 91–99. Springer, Berlin, Heidelberg (2010)
8. Kolenikov, S., Angeles, G.: The use of discrete data in PCA: theory, simulations, and applications to socioeconomic indices. Chapel Hill: Carolina Population Center, University of North Carolina **20**, 1–59 (2004)

Chapter 8
Discussion

Abstract In case of reliability checking, more focus should be given on the questionnaire that aims to collect demographic information, facts, attitudes and personal options from respondents. The questionnaire should contain the ability of thoughts on a particular subject matter. During this period of uncertainty, it was indeed a real challenge to collect a large set of data. This section is focused on the estimation of sample size, challenges of data collection, dealing with small datasets and also highlights the importance of validity and reliability of the questionnaires.

8.1 Sample Size and Estimation of Sample Size

A smaller set of data is represented by a sample. The researchers select a sample from a larger population by using a proper predetermined selection method. Sampling is the most efficient method of conducting research. To study the entire population is time-consuming, costly, and impossible also, that is why it is better to examine a sample that is drawn from the population. The characteristics of the population should exist in the sample as well. The process by which a sample can be drawn from a population is known as the sampling method. The sampling method contains qualitative data and quantitative data. It is an integral part of research design. Sampling methods are categorized in two ways (i) probability sampling and (ii) non-probability sampling. Probability sampling is a method where the population is based on the theory of probability. This method is including everyone in the population. Everyone has an equal chance to be selected in this sampling method. In this type of sampling method, no biasness is entertained. Under the probability sampling method, a very important technique is simple random sampling.

In our study, a random sample (each member has an equal chance to be selected) is taken from the population for making statistical interference and studied in deep to identify the characteristics of the entire population. This includes the mean or measure of central tendency, the standard deviation or measure of spread, and the proportions of sample units in the population. In a collection of numbers mean is the most common value, it means the average of values. In the probability distribution of the sample, it is a measure of central tendency along with the concept of median

G. R. Shinde et al., *Quality of Work-Life During Pandemic*, Studies in Big Data 100,
https://doi.org/10.1007/978-981-16-7523-2_8

and mode. In finance by using the statistical concept mean, a major significance can be drawn. On the other hand, the standard deviation refers to the measurement of how dispersed the data is toward the mean. If there is a low standard deviation, it signifies that data is clustered around the mean. High standard deviation signifies data are more spread out. There is a need to study sample distribution for understanding the distribution of outcomes [1, 2]. It is very much interesting to observe different statistics that have various sampling distributions. The distribution shapes depend on (i) the specific statistic, (ii) the sample size and (iii) the parent distribution. In a sampling distribution, the variability can be decreased by enhancing the sample size. With the help of sample size, we can understand a particular group of subjects that have been selected from the general population size. It is representative of the real population. If the sample size is greater than necessary, it is a waste of resources. On the other hand, very small sample size does not provide the accurate result or the right decisions. The main aim of the study is to obtain the desired level of accuracy along with that desirable confidence level at the minimum cost. In the field of statistics, the confidence level refers to the probability, an estimation of the location of a statistical parameter. In surveys, most of the time the level of confidence has been used as 90 or 95 or 99%. The margin of error signifies the statistical expression regarding the amount of random sampling error in the results of a survey. The larger margin of error is generating less confidence level in the result of a survey of the entire population. The two important components like margin of error and sample size are inversely proportional. In our study, we have used the raosoft formula to estimate the sample size for each sector. There is a requirement for a higher confidence level in case of a larger sample size. It is established in the formula that for a population larger than 20,000 sample size does not change much. Response distribution should be 50%.On the other hand, through this formula, it can be said larger sample size can be generated if a lower margin of error exists. After a certain point, if the sample size is increased beyond the provided sample, it generates a diminished return because the increase in accuracy will be very negligible. The confidence level signifies the degree of confidence that includes the data which is obtained from the sample [3, 4]. Population characteristics are reflected in the sample size that has been chosen through the random sampling technique. For a population larger than 20, 000, there is no such change in sample size. The distribution of response refers to the expected results for each question. The sample size n and margin of error E are represented by:

$$X = Z(c/100)^2 r(100 - r)$$

$$N = Nx/((N - 1)E^2 + x$$

$$E = \sqrt{[(N - n)x/n(N - 1)]}$$

where N is the population size, r is the fraction of responses that you are interested in, and $Z(c/100)$ is the critical value for the confidence level c.

The formula to estimate sample size is explained in this way, where the acceptable margin of error is 5% and the level of confidence is 95%, then the size of the population is 20,000. Response distribution is 50% as there is a chance of 50–50 to get the answer from respondents. Therefore, the recommended sample size is 377 (as per http://www.raosoft.com/samplesize.html calculator). For selecting a sample of the respondent population identification is very important. In a sampling distribution, the variability can be reduced by increasing the sample size.

8.2 Data Collection Challenges

In this study, we have collected data from different sectors in India, namely education, real estate, agriculture, etc. All these sectors are badly impacted by the pandemic. Fear of getting infected, prolonged quarantine duration, frustration, fatigue, monotony, boredom all are generated during this crisis time. Apart from this financial loss, mental stigma, misinterpreted news, etc., have created a wide range of symptoms of psychological stress and disorder. Collecting data from these sectors was a real challenge in this uncertain time. In most cases, there are no such positive responses. Sometimes people are not willing to give respond due to fear of expose. In any questionnaire-based survey, a thin line of defense relying on the honesty of the participants, who answer the questions truthfully, always exists.

8.3 Problems with Less Sample Size

There are numerous problems of small sample size, but they mostly revolve around large variance [5]. For small datasets, overfitting becomes more difficult to avoid. In cases of machine learning model design with those data, most of the time not only is the training data overfitted but sometimes the validation set is also overfitted. Outliers become much more dangerous [6, 7]. Lastly, noise often becomes a real problem, either in its target variable or in certain functions.

8.4 Validity and Reliability of Questionnaires

For considering a questionnaire to be accepted, we must do two very important parameters to check, one is validity another one is reliability. The first one measures the consistency of a questionnaire and the second one measures the degree to which we can agree to the results from the questionnaire with the real world. To check the validity of the questionnaire, we have done the pilot test. In this study, we enter the data into a spreadsheet and clean the data. One person reads the values, and another person enters the values into a sheet. This helps to reduce mistakes while entering data [8]. If all the questions are answered by respondents correctly, then their responses to the negative paraphrased questions will match with similar positively phrased questions. If any inconsistency is incorporated, then the person's questionnaire must be tossed out [9]. We should check the minimum value as well as maximum for the entire dataset. From a 5-point Likert scale if any responses come at 6, the error should be identified in this case.

In case of reliability checking, we should focus on the questionnaire that aims to collect demographic information, facts, attitudes, personal options from respondents. The questionnaire should contain the ability of thoughts on a particular subject matter. In this study, we have focused on the quality of the work-life of different sectors through an appropriate questionnaire. The design of the questionnaire directly affects the type of information that is required from the questionnaire [10, 11]. Qualitative questionnaires are applied for gathering expository information while quantitative questionnaires are used to validate the previously generated hypothesis.

8.5 Summary

During this period of uncertainty, it was indeed a real challenge to collect a large set of data. We understand that for more general modeling, a greater number of responses are required besides validity and reliability testing. These can be the future scope of our study.

References

1. Israel, G. D. (1992). Determining sample size.
2. Krejcie, R.V., Morgan, D.W.: Determining sample size for research activities. Educ. Psychol. Measur. **30**(3), 607–610 (1970)
3. Morse, J. M. (2000). Determining sample size.
4. MacCallum, R.C., Widaman, K.F., Zhang, S., Hong, S.: Sample size in factor analysis. Psychol. Methods **4**(1), 84 (1999)
5. Kirk, J., Miller, M. L., & Miller, M. L. (1986). Reliability and validity in qualitative research (Vol. 1). Sage.
6. Hand, D. J., Daly, F., McConway, K., Lunn, D., & Ostrowski, E. (1993). A handbook of small data sets. cRc Press.
7. Martens, H.A., Dardenne, P.: Validation and verification of regression in small data sets. Chemom. Intell. Lab. Syst. **44**(1–2), 99–121 (1998)
8. Desu, M.M.: Sample size methodology. Elsevier (2012)
9. Roberts, P., Priest, H.: Reliability and validity in research. Nurs. Stand. **20**(44), 41–46 (2006)
10. Fink, A., & Litwin, M. S. (1995). How to measure survey reliability and validity (Vol. 7). Sage.
11. Golafshani, N.: Understanding reliability and validity in qualitative research. The qualitative report **8**(4), 597–607 (2003)

Chapter 9
Conclusions

Abstract This book explains the quality of work-life, its characteristics and stages along with the different factors causing its importance. The effect of pandemic on the employees of real estate, education, software industries and agriculture sectors are also reported using data analysis and various mathematical modeling. It is important to note that to maintain a good and effective quality of work-life, not only employees are responsible, but also the employers should extend their hand of help toward the employees. They must provide all necessary wellness facilities and upgrades that are needed at this crisis time.

9.1 Conclusions

Quality of work-life is a prime practice in any organization under the vast field of human resource management. This practice gives the organization an effective shape in terms of environment, methodologies and outcomes. The quality of work-life has twin goals to accomplish; it increases the effectiveness of the functions of organizations. On the other hand, it helps to improve the quality of life of the working people at work. In generic terms, it can be said that quality of work-life is covering the area of people's feelings toward the dimensions of work in the work environment. If we focus on one of the important parameters of qualitative work-life, we must consider work-life balance. At present time, due to the drastic effects of the COVID-19 pandemic the quality of work-life of the working people is getting poor day by day. Employees are far away from the workplace and are doing remote work in this scenario. All the time, all kinds of facilities and resources are not available to do work through this man–machine interface. The work from home concept makes the work-life balance unbalanced. Working people are not able to manage professional life as well as personal life during these phases. As lots of work is coming to them from their workplace as well as from home too. The work-life balance indicates the picture where the employees can manage their professional life as well as personal life too. The work from home has both sides, and it becomes easier to work due to the availability of cloud-based software, mobile technology, advantages of smartphones and prolong activities of the Internet, and the other side is it may increase a stress

G. R. Shinde et al., *Quality of Work-Life During Pandemic*, Studies in Big Data 100,
https://doi.org/10.1007/978-981-16-7523-2_9

level of an employee. It can be said that by using these technologies, the business entrepreneurs can run their operations easily but on the other hand, the excessive level of stress makes the employee's life blur. It becomes such a practice where employees need to produce more in less time, the result is physical and mental health problems are raised. In order to maintain a good and effective quality of work-life, not only the employees are responsible, but also the employers should extend their hand of help toward the employees. They should provide all the welfare facilities and betterment that are needed at this time. Work-life balance and mental health considerations are very important as per the guidelines issued by WHO during this COVID-19 outbreak.

In view of this, an exploratory study is required to investigate the ongoing experiences of employees and employers faced in India. A question-based survey for employees in different sectors like education, information technology, agriculture, health care and real estate sector is required to collect their experiences on the different types of questions. On the responses received for these questions, mathematical tools can play an important role to build mathematical models for employees in different sectors. The application of appropriate tools from mathematics and mathematical model depends on the set of a questionnaire prepared for different domains. In addition to this, machine learning-based analysis is also important to draw inference from the outcomes resulted from clustering or classification algorithms in machine learning.

The first part of this book explains the quality of work-life, its characteristics and stages along with the different factors causing its importance. Methodologies to improve the quality of work-life are also explained in detail in the first chapter of the book. Mathematical modeling based on data analysis plays important role in predictions and forecasting. The second chapter of the book deals with the need for mathematical modeling, various tools and languages in mathematics. Pragmatic approaches for mathematical modeling based on rough sets, conditional probability and graph theory are explained in detail in this part.

The second part of the book covers specific case studies in various sectors which include real estate, education, software industries and agriculture. This part presents and discusses an overview of every sector, the study of state of the art, descriptive statistics of the dataset for qualitative and qualitative analyses. Mathematical modeling and machine learning-based analysis of the underlined dataset are also presented in this part of the book. The case study of the real estate sector is discussed in the third chapter of the book. More than 90% of respondents of general/supervisor level from the real estate sector disagree that the company is taking care of their wellness including mental health during this pandemic. 100% agree responses regarding health and wellbeing came from the middle level only. All responses of strongly agree and strongly disagree for health and wellbeing are from top level and middle level, respectively.

In the fourth chapter, the education sector case study is discussed. In the educational sector, 0.6% of respondents from age group 20–29 are averagely impacted on their mental health and overcome the family crisis. Respondents having more than 50 years of age and the pandemic impacted very high on their mental health and wellbeing. Respondents from less than 30 years or more than 50 years of age feel that there is no family crisis during a pandemic. It is observed that during a pandemic, job security hamper in the educational sector. Every respondent concedes that there is a notable impact of a pandemic on salaries. In this period, there is less opportunity for Principles/Directors/Deans to shift their jobs. That is, high salaried respondent does not have more job opportunities during a pandemic. Every respondent has managed good work–life balance during this COVID-19 pandemics, but Principles/Directors/Deans/Professors have balanced more work-life balance than others.

The IT/software industry case study is discussed in the fifth chapter. Rough set theory is applied to the data of the IT sector, and the decision rules are designed to know the effect of COVID-19 on mental health during the pandemic period. It is found that if the job security is low and additional support by the employer during the pandemic is moderate then COVID-19 is certainly affected the mental health and wellbeing of the respondents from the IT sector. Moreover, if the work satisfaction is low and work-life balance is low during a pandemic, then COVID-19 is certainly affected mental health and wellbeing. In this sector, respondents feel that there is a low work-life balance in this pandemic.

In the sixth chapter, case study of the agriculture sector is discussed. We have designed decision-making rules to know the impact of COVID-19 on mental health and wellbeing during the pandemic period for the respondents from the agricultural sector by using rough set theory. In the agricultural sector, if someone is scared more of getting infected during a pandemic, then COVID-19 is affected on the mental health certainly during the pandemic period. If the respondents are satisfied with the protective and preventive measures taken by the government, then COVID-19 is not affected on the mental health certainly during the pandemic period.

The last part of the book addresses the need for principal component analysis (PCA) for categorical dataset generated from the survey. In Chap. 7, an overview of PCA, computational stages and its interpretation are elucidated. In addition to this, the size of the sample dataset, the margin of error, hypothesis testing, the challenge in data collection, analysis of the small data are also presented and discussed in Chap. 8.

This uncertain spread of COVID-19 is creating emotional challenge for many people leading to their daily routine in very unprecedented ways. Every stakeholder of the society like employer, employees is being affected to large extent and is responsible to play important role in controlling this spread. Preventing and mitigating COVID-19 at workplace are very important and require proper guidelines and recommendations to prevent and mitigate COVID-19 at workplace. Work-life balance and mental health considerations are very important as per the guidelines issued by World Health Organization (WHO) during this COVID-19 outbreak. As per the guidelines stated by WHO, the recommendations are issues to the different groups of people which include general population, healthcare workers, team leaders in healthcare organizations, older adults, people with underlying health conditions and their careers, people in isolation, etc.

Due to this pandemic, work from home culture has become policy priority for most state and central governments and there should be policy in place considering both employers and employees. In the view of this, an exploratory study is required to investigate the on-going experiences of employees and employers faced in India. Question-based survey for employees in different sectors like education, information technology, agriculture, healthcare and real estate sector is required to collect their experiences on the different types of questions. On the responses received for these questions, mathematical tools can play an important role to build the mathematical models for employees in different sectors. The application of appropriate tool from mathematics and mathematical model depends on the set of questionnaire prepared for different domains. In addition to this, machine learning-based analysis is also important to draw inference from the outcomes resulted from clustering or classification algorithms in machine learning. The normative inferences are also one of the important outcomes of these machines learning-based inferences to decide what is ought to be? The selection of appropriate machine learning algorithm will completely depend on the descriptive statistics of the dataset, and accordingly, the outcomes can be analyzed for different purposes. The key contributions presented in the scope of this book are listed as follows:

1. The preparation of questionnaire for employees in different sectors for recording their ongoing experiences during this COVID-19 pandemic.
2. Preparation of the dataset resulted from the responses collected from the survey initiated for employees in different sectors.
3. Proposed sector-specific mathematical model to draw statistical inferences, correlation in the form of association rule mining and key observations.
4. Proposed machine learning-based analysis to draw useful inferences and normative interpretations from the dataset.
5. Set if useful recommendations and guidelines.

Index

G. R. Shinde et al., *Quality of Work-Life During Pandemic*, Studies in Big Data 100,
https://doi.org/10.1007/978-981-16-7523-2